67 Topics in Current Chemistry
Fortschritte der chemischen Forschung

Preparative Organic Chemistry

Springer-Verlag
Berlin Heidelberg GmbH 1976

This series presents critical reviews of the present position and future trends in modern chemical research. It is adressed to all research and industrial chemists who wish to keep abreast of advances in their subject.

As a rule, contributions are specially commissioned. The editors and publishers will, however, always be pleased to receive suggestions and supplementary information. Papers are accepted for "Topics in Current Chemistry" in either German or English.

ISBN 978-3-662-15847-0   ISBN 978-3-540-38041-2 (eBook)
DOI 10.1007/978-3-540-38041-2

© by Springer-Verlag Berlin Heidelberg 1976
Originally published by Springer-Verlag Berlin Heidelberg New York in 1976
Softcover reprint of the hardcover 1st edition 1976

The use of registered names, trademarks, etc. in this publication does not imply, even in the absence of a specific statement, that such names are exempt from the relevant protective laws and regulations and therefore free for general use.

Typesetting and printing: Schwetzinger Verlagsdruckerei GmbH, 6830 Schwetzingen. Bookbinding: Konrad Triltsch, Graphischer Betrieb, Würzburg

Contents

Old and New in the Field of Directed Aldol Condensations

Professor Dr. Georg Wittig

Institut für Organische Chemie der Universität Heidelberg, Im Neuenheimer Feld 270,
D-6900 Heidelberg, Germany

This paper was presented as a lecture at the meeting of the Gesellschaft Deutscher Chemiker in Köln on September 10, 1975.

Contents

1. Introduction

In connection with the investigation of 1,2-dehydrobenzene, phenyllithium was reacted with fluorobenzene in the presence of diethylamine[1]. As expected diethylaniline was among the reaction products, resulting from the addition of diethylamine to dehydrobenzene, as well as α-phenyldiethylamine IV. A more detailed investigation of these results showed that the lithiumdiethylamide which developed during the reaction also added to the intermediate dehydrobenzene forming the non-isolable ate-complex I. The anionic loosening of ligands in the neighbourhood of nitrogen causes a hydride transfer which results in the formation of phenyllithium and Schiff base II. These compounds combine irreversibly by addition of the organolithium to the C=N-bond producing lithiumamide III, which following hydrolysis yields α-phenyl-diethyl-amine IV.

In order to establish the reducing agent function of the dialkylamide it was reacted with benzophenone[2]. We expected a reduction by way of the ate-complex V

2

to benzhydrol coupled with the oxidation of the amine to Schiff base II. Although following hydrolysis, approximately 35% yield of benzhydrol was isolated; a complicated compound was obtained instead of ethylidenethylamine II. A meticulous structural determination and a later synthesis showed the unexpected compound to be the Schiff base VII, *i.e.* the aldimine adduct with benzophenone.

Evidently a fraction of the benzophenone is reduced to benzhydrol in a redox reaction in which an equivalent amount of Schiff base II is formed. After metallation of II by the remaining lithiumdiethylamide, the metallated Schiff base VI reacts with the reduced benzophenone to yield the aldimine adduct VII. Surprisingly, the entire process is completed within one minute at 0 °C.

Since the final product can be easily converted with a mineral acid to β-phenyl-cinnamaldehyde, the total reaction is an aldol condensation. The methyl group of the acetaldehyde is specifically condensed with the carbonyl group of the ketone. This opportunity of achieving a directed aldol condensation was now put into practice and the method was quickly shown to be capable of a wide application with generally high yields.

Previously, it was not possible to control the aldol condensation so that aldehydes would combine with the carbonyl groups of ketones with their α—CH-groups. For example, acetaldehyde does not react with benzophenone under the usual base catalyzed conditions to give adduct VIII because the aldehyde undergoes a much more rapid self addition to acetaldol IX.

Even the reaction with more active ketones, such as acetone, does not lead to adduct X but, as has been proven, to compound XI. Acetaldehyde has the more easily accessible carbonyl group and acetone, following anionization, possesses the more strongly nucleophilic α-carbon.

$$(C_6H_5)_2C-CH_2-CH \xleftarrow{\text{Base}}_{\text{//}} (C_6H_5)_2C + CH_3-CH \xrightarrow{\text{Base}} (C_6H_5)_2C + CH_3-CH-CH_2-CH$$

VIII   IX

X   XI

2. Directed Aldol Condensation

2.1. General Procedure for Directed Aldol Condensation

As the aldol condensation described at the outset has clearly shown, these difficulties can be avoided if the aldehyde carbonyl group is converted to an azomethine group.

This Schiff base is metallated with an appropriate lithium compound to XII and subsequently reacted with benzophenone or acetone. Since the protective group can be removed following the aldol addition, this method is a means of bringing about organometallic directed aldol condensations[3].

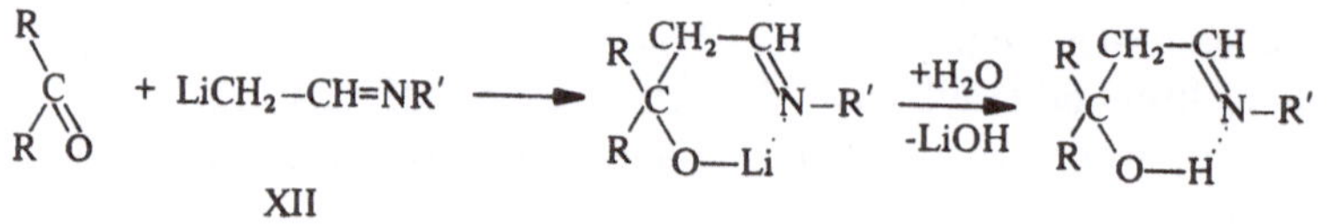

$$R=CH_3 \text{ or } C_6H_5$$

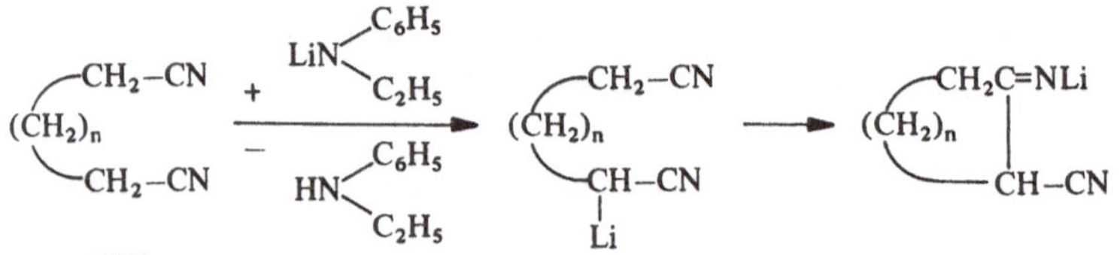

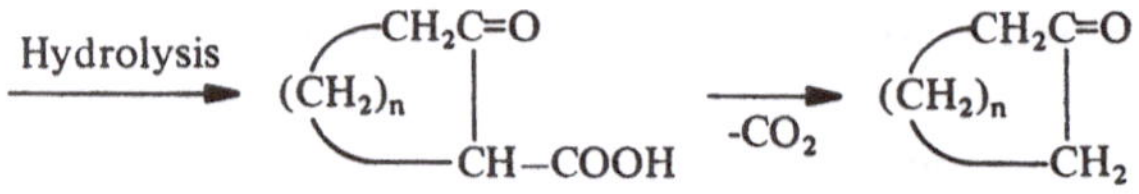

In this connection it should be remembered that Karl Ziegler[4] had already used metallated secondary amines for condensation purposes. For the synthesis of larger ring compounds he applied the lithium ethylanilide in a type of Dieckmann reaction to ring close dinitriles, such as XIII. Since the metallated amine is soluble in ether, he succeeded in forcing the cyclization by the principle of dilution.

The resulting product led to cyclic ketones such as Exaltone (Cyclopentadecanone) or Dihydrocivetone (Cycloheptadecanone) in yields of 60—70%, after hydrolysis with saponification of the ketimine group and cleavage of CO_2. What an ingenious move in a successful operation!

Returning to our problem, we determined the following optimal conditions under which the metallation of the Schiff base and the succeeding reaction with carbonyl compounds, to give the aldimine adducts, should proceed. Cyclohexylamine has proved to be effective as the amine component of the Schiff base, because Schiff bases with branched alkyl substituents on nitrogen have a lesser tendency of self-addition than those with unbranched groups. Further, it was shown, also for steric reasons, that lithium diisopropylamide was a more appropriate metallating agent than lithium diethylamide. When accordingly, ethylidene cyclohexylamine was reacted with lithium diisopropylamide in ether at 0 °C and after 10 minutes benzophenone was added at −70 °C, the Schiff base XIV could be recovered in 92% yield following hydrolysis[5].

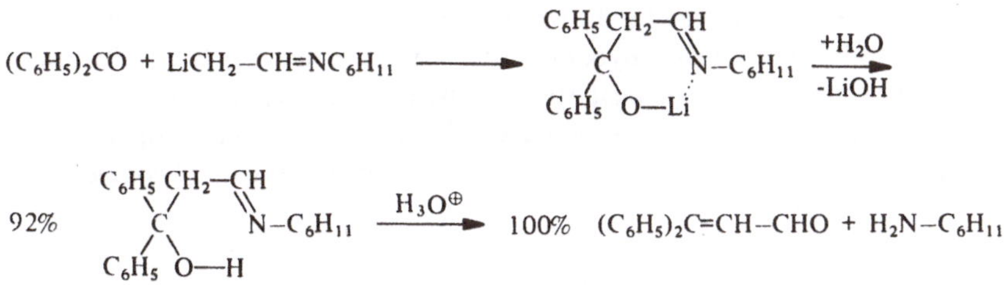

$$(C_6H_5)_2CO + LiCH_2-CH=NC_6H_{11} \longrightarrow \underset{C_6H_5 \quad O-Li}{\overset{C_6H_5 \; CH_2-CH}{C}} \underset{}{N-C_6H_{11}} \xrightarrow[-LiOH]{+H_2O}$$

$$92\% \quad \underset{C_6H_5 \quad O-H}{\overset{C_6H_5 \; CH_2-CH}{C}} N-C_6H_{11} \xrightarrow{H_3O^{\oplus}} 100\% \; (C_6H_5)_2C=CH-CHO + H_2N-C_6H_{11}$$

XIV

These favourable results can probably be traced back to the stable chelates formed by the lithium salts of the aldol adducts. In the hydrolysis product the nitrogen of the aldimine group is linked through an intramolecular hydrogen bond bridge, as was established by the IR spectrum. Since the aldimine adduct could finally be converted by treatment with acid in a nearly quantitative yield to β-phenylcinnamaldehyde, the preparative problem of subjecting aromatic ketones to the directed aldol condensation was solved.

However, even in the singularly smoothly operating directed aldol condensation with aromatic ketones, there is no such thing as perfection. In order to determine the influence of inductive and steric effects on the new process, the Schiff bases of α-substituted acetaldehydes were metallated under the standard conditions with lithium diisopropylamide, and the aldimine adducts with benzophenone were isolated. Table 1[6] shows the measured yields.

Table 1

Schiffbase	Aldimine-adduct	m. p.
$CH_3-CH=NR$	92%	127–128°
$\overset{CH_3}{\underset{\|}{CH_2}}-CH=NR$	91%	82–83°
$\overset{C_2H_5}{\underset{\|}{CH_2}}-CH=NR$	71%	104–105°
$\overset{CH_3}{\underset{\|}{HC}}-CH=NR$ $\underset{CH_3}{\|}$	29%	107–108°
$\overset{C_2H_5}{\underset{\|}{HC}}-CH=NR$ $\underset{C_2H_5 \quad XV}{\|}$	0%	

According to these values, the yields decrease with increasing branching on the α-carbon of the Schiff base until they reach 0% with 2,2-diethyl-ethylidene cyclohexylamine XV.

5

The basis for these findings can be shown by the following: When the Schiff base XV was treated with lithium diisopropylamide for two hours in ether (the Schiff base of the unsubstituted acetaldehyde is instantly metallated in ether) and then reacted with D_2O, the starting material XV was recovered almost quantitatively. Since the compound does not contain any deuterium, as shown by NMR, it can be concluded that it was not metallated.

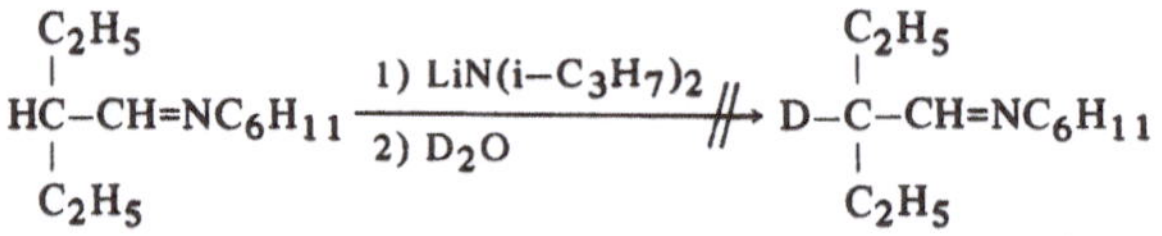

$$\begin{array}{c} C_2H_5 \\ | \\ HC-CH=NC_6H_{11} \\ | \\ C_2H_5 \end{array} \quad \xrightarrow[\text{2) } D_2O]{\text{1) } LiN(i-C_3H_7)_2} \quad \not\longrightarrow \quad \begin{array}{c} C_2H_5 \\ | \\ D-C-CH=NC_6H_{11} \\ | \\ C_2H_5 \end{array}$$

XV

The increasing alkylation of the aldimine leads to a reduced suitability for being metallated. This is likely to be explained by the interaction of the growing steric hindrance and the hyperconjugation effect which lowers the mobility of the α-proton.

The Schiff bases of ketones can also be condensed with aromatic ketones, such as benzophenone, by the method under discussion[7]. Using the standard conditions, isopropylidene cyclohexylamine was metallated to XVI and combined with benzophenone. The ketimine adduct XVII could be isolated in 65% yield. Following the decomposition of XVII with diluted sulphuric acid the unsaturated ketone XVIII was recovered. It proved to be identical with a comparable compound which was synthesized by a more cumbersome procedure. The nitrogen free ketol could also be obtained from XVII. During thin layer chromatography of XVII on silica gel, benzo-

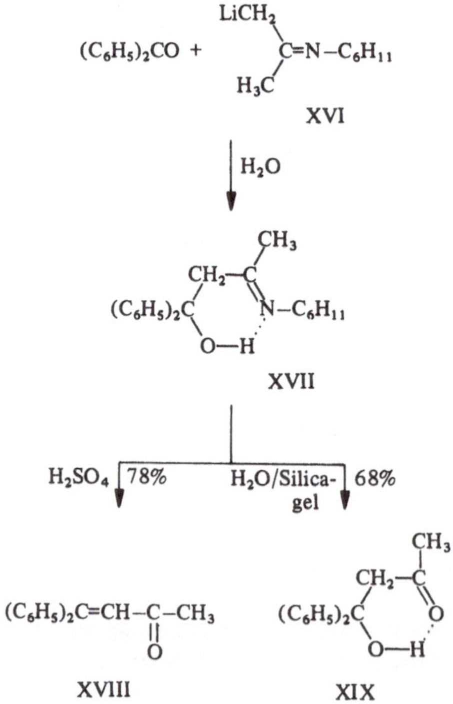

$$(C_6H_5)_2CO \; + \; \begin{array}{c} LiCH_2 \\ \diagdown \\ C=N-C_6H_{11} \\ \diagup \\ H_3C \end{array}$$

XVI

$$\downarrow H_2O$$

$$\begin{array}{c} CH_3 \\ | \\ CH_2-C \\ (C_6H_5)_2C \qquad \diagdown N-C_6H_{11} \\ \diagdown \\ O-H \end{array}$$

XVII

$$H_2SO_4 \boxed{78\%} \qquad H_2O/\text{Silica-}\boxed{68\%}$$
$$\text{gel}$$

$$(C_6H_5)_2C=CH-\underset{\underset{O}{\|}}{C}-CH_3 \qquad \begin{array}{c} CH_3 \\ | \\ CH_2-C \\ (C_6H_5)_2C \qquad \diagdown O \\ \diagdown \\ O-H \end{array}$$

XVIII   XIX

phenone was found as well as a compound with a shorter path of travel. This substance had a melting point of 85 °C and the elemental formula $C_{16}H_{16}O_2$. On the basis of the NMR spectrum and the proton ratios ($CH_{aromatic} : OH : CH : CH_3 = 10 : 1 : 2 : 3$), it is the previously unknown ketol XIX.

By reaction of the Schiff base of the homologous diethylketone with benzophenone, the ketimine adduct XX could indeed be isolated in 60% yield, but the respective ketol could not be obtained by thin layer chromatography. The same lack of success resulted from the attempted acid decomposition of XX to the corresponding unsaturated ketone. In both cases only benzophenone was recovered because here the branching from the α-carbon atom also counteracts the stability of the ketol adduct.

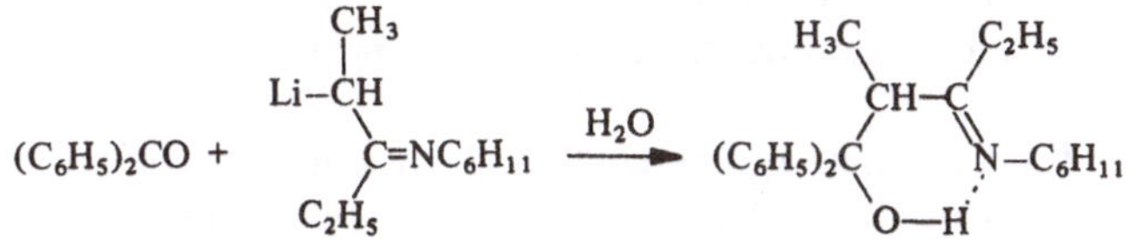

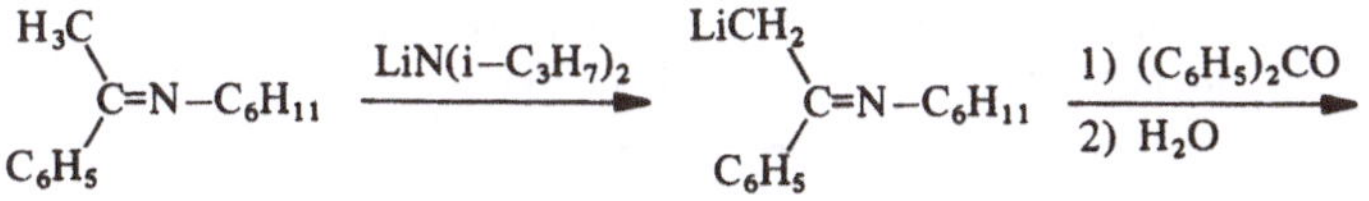

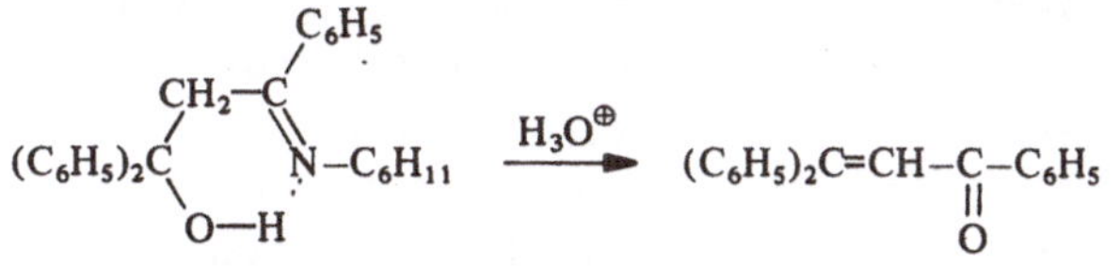

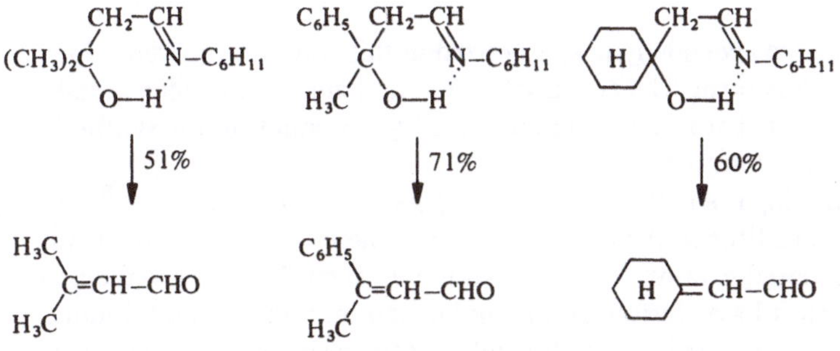

The Schiff base of acetophenone XXI, after anionization with lithium diisopropylamide to the deep yellow solution of XXII, led to the adduct XXIII (55% yield) upon combination with benzophenone. Treatment with diluted mineral acid converted XXIII in 95% yield to 1,1-diphenyl-2-benzoylethylene[6].

7

At first this new technique of adding metallated Schiff bases appeared problematical for the possible synthesis of natural products, because during its application to methylketones, such as β-ionone, it could be expected that the proton active $-COCH_3$ system would undergo a transmetallation or transanionization with the metallated Schiff base. Fortunately, this was not the case[3], as was shown by model experiments with acetone, acetophenone, and cyclohexanone.

2.2. Application to the Synthesis of Natural Products

In general, the aldimine adducts of natural products can be converted in one step to the corresponding α, β-unsaturated aldehydes or ketones. They are subjected to a steam distillation in the presence of oxalic acid, in which the final product distils over. Under these conditions the more thermodynamically stabile product is formed, usually the trans olefin.

The compounds listed in Table 2 can be synthesized relatively simply by the method under discussion[7]. The yields given relate to the starting aldehyde or ketone.

Table 2

Carbonyl component	Aldimine adduct m. p. and yield	α, β-Unsaturated aldehyde b.p. and yield
Butyraldehyde		2-Hexene-1-al, (Leafaldehyde) b.p._{15} 42–48 °C, 65%
β-Cyclocitral	92–93 °C 58%	β-Cyclocitrylidene-acetaldehyde b.p._{0.1} 72.5–75 °C, 50%
6-Methyl-5-heptene-2-one	32–33.5 °C 76%	Citral (*cis-trans*-mixture) b.p._{0.5} 68–74 °C, 64%
β-Ionone	46–47 °C 80%	β-Ionylidene-acetaldehyde b.p._{0.001} 115–120 °C, 42%

The β-cyclocitrylidene acetaldehyde synthesized in this way was recovered for the first time in crystalline form. The two aldehydes, citral and β-ionylidene acetaldehyde, are of special value because they are intermediate products in the synthesis of Vitamin A and other carotinoids.

Also worth mentioning is a natural product synthesis performed by Büchi[8]. It concerns β-sinensal, a constituent of the aroma of the chinese orange. Starting from myrcene, the bromination derivative was prepared in two steps. This was condensed with the metallated Schiff base, derived from ethylidene-t-butylimide and lithium diisopropylamide, and hydrolyzed to the aldehyde in 60% yield with aqueous oxalic acid. The resulting aldehyde was now reacted with metallated propylidene-t-butylimide, in the same way as the directed aldol condensation, and again hydrolyzed with aqueous oxalic acid (60% yield). The final product was identical in all properties with the naturally occurring β-sinensal. Through chemical modification of the intermediate product it was proven, that the sesquiterpene has the trans-trans configuration.

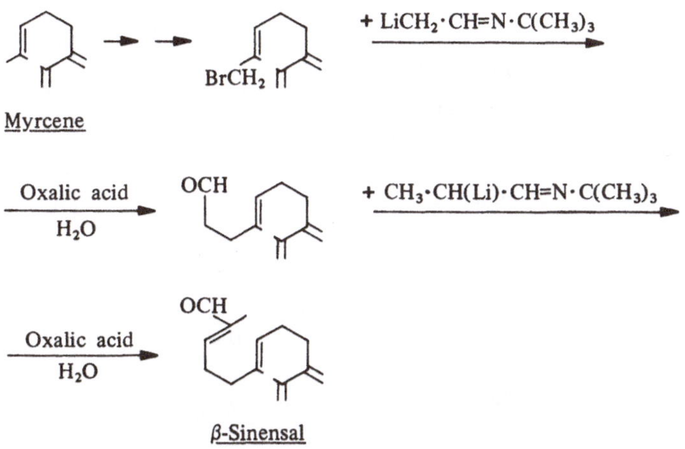

$+ \text{LiCH}_2 \cdot \text{CH=N} \cdot \text{C(CH}_3)_3$

BrCH$_2$

Myrcene

$\dfrac{\text{Oxalic acid}}{\text{H}_2\text{O}}$ OCH

$+ \text{CH}_3 \cdot \text{CH(Li)} \cdot \text{CH=N} \cdot \text{C(CH}_3)_3$

$\dfrac{\text{Oxalic acid}}{\text{H}_2\text{O}}$ OCH

β-Sinensal

3. Directed Aldol Condensation and Modified Carbonyl-Olefination with Phosphorus Ylids — a Comparison

The essential point of the directed aldol condensation[9] is that it makes a desired extension of the chain of an aldehyde or ketone to an α,β-unsaturated ketone or aldehyde possible. It could be argued that such a chain extension is also possible by means of an olefination using phosphorus ylids. Certainly, this latter method has a broad range of application and it was used by Trippett[10] to obtain α,β-unsaturated aldehydes similar to those under discussion. It has an added advantage in that the dehydration of the carbonyl-alcohol resulting from the directed aldol condensation, which does not always progress smoothly, can be avoided.

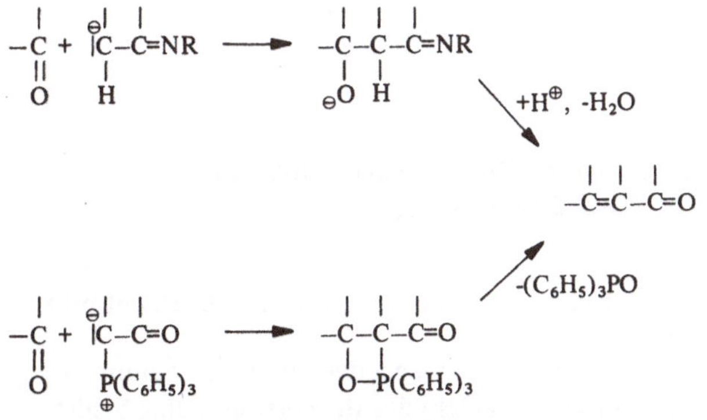

$+\text{H}^{\oplus}, -\text{H}_2\text{O}$

$-(\text{C}_6\text{H}_5)_3\text{PO}$

However, the olefination by way of the phosphorus ylid also has its disadvantages particularly in this sphere. The scope of both methods is compared in Table 3[9] in which the yields, as before, are based on the starting aldehyde or ketone.

It is readily apparent that the olefination through the phosphorus ylid gives good results with aldehydes as the substrate. On the other hand, it fails utterly when ap-

Table 3

Final product	Prepared by	
	directed aldol condensation	phosphorus ylids
$CH_3-CH_2-CH_2-CH=CH-CHO$	65%	81%
$C_6H_5-CH=CH-CHO$	72%	60%
$(CH_3)_2C=CH-CHO$	40%	0%
$\begin{array}{l}C_6H_5 \\ \diagdown \\ C=CH-CHO \\ \diagup \\ CH_3\end{array}$	71%	0%
$(C_6H_5)_2C=CH-CHO$	78%	0%
$\begin{array}{c}CH_3 \\ \mid \\ (C_6H_5)_2C=C-CHO\end{array}$	81%	0%
$\begin{array}{c}CH_3 \\ \mid \\ (C_6H_5)_2C=CH-C=O\end{array}$	50%	0%

plied to ketones. It is just these condensations with ketones which are made possible by the directed aldol condensation using metallated Schiff bases. It thus finds its proper sphere of application.

The reason for the failure of the carbonyl-olefination with ketones by way of the phosphorus ylid lies in the resonance stabilization of the ylid. The charge distribution is concentrated in the direction of the oxygen atom. Therefore, it also appeared fitting here to introduce a C=N–R group in place of the C=O group in phosphorane.

$$-\overset{|}{\underset{\overset{\|}{O}}{C}} + \overset{\ominus}{\underset{\oplus P(C_6H_5)_3}{\overset{|}{C}-\overset{|}{C}=N \cdot R}} \longrightarrow -\overset{|}{C}-\overset{|}{\underset{\overset{|}{O}-P(C_6H_5)_3}{C}}-\overset{|}{C}=N \cdot R \xrightarrow[-(C_6H_5)_3PO]{} -\overset{|}{C}=\overset{|}{C}-\overset{|}{C}=N \cdot R$$

4. Combination of the Directed Aldol Condensation and the Carbonyl-Olefination with Phosphorus Ylids

This idea, published by us in 1968[9], has in the meantime (in 1969), been put into practice by Japanese scientists[11].

The phosphonate XXIV was converted to the respective ylid with sodium hydride and yielded on reaction with ketones or aldehydes the corresponding Schiff bases XXV. These products, in contrast to the aldehyde or ketimine adducts with their intramolecular hydrogen bonds, can easily be converted to the unsaturated carbonyl compound in good yield.

Since the realization of our ideas concerning the triphenylphosphoranes has only recently begun and the experimental work is not yet finished, I can only give a short preliminary report on the subject[12].

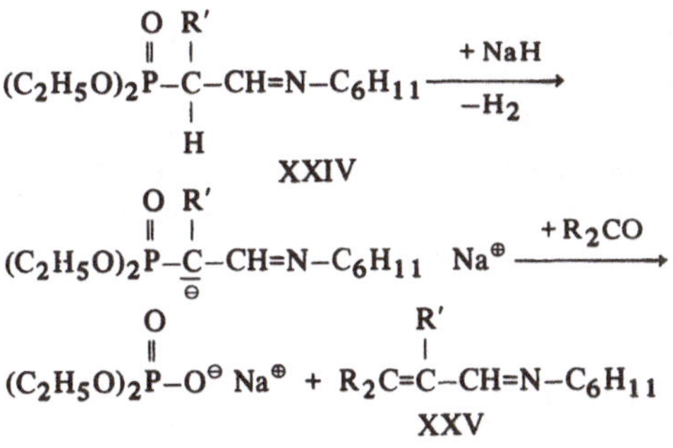

$$\underset{\underset{H}{|}}{\overset{\overset{O}{\|}\ \overset{R'}{|}}{(C_2H_5O)_2P-C-CH=N-C_6H_{11}}} \xrightarrow[-H_2]{+NaH}$$

$$\text{XXIV}$$

$$\underset{\overset{|}{\ominus}}{\overset{\overset{O}{\|}\ \overset{R'}{|}}{(C_2H_5O)_2P-C-CH=N-C_6H_{11}}}\ Na^{\oplus} \xrightarrow{+R_2CO}$$

$$\overset{\overset{O}{\|}}{(C_2H_5O)_2P-O^{\ominus}}\ Na^{\oplus}\ +\ R_2C=\overset{\overset{R'}{|}}{C}-CH=N-C_6H_{11}$$

$$\text{XXV}$$

We reacted the phosphonium salts XXVIa–XXVIc[a] with carbonyl compounds under varied experimental conditions. It was shown that when the olefination was carried out in absolute ether with lithium diisopropylamide as deprotonation agent – in accordance with the reaction under discussion – the phosphonium salt XXVIa and benzaldehyde formed the Schiff base of the cinnamaldehyde XXVII in 77% yield. In comparison the yield is reduced when, in place of the p-N,N-dimethylamino-phenyl residue in XXVIa, the p-methoxyphenyl residue (XXVIb) is introduced. Using approximately the same reaction conditions, it falls to 65% of XXVIII and still further to 36% of XXIX when the cyclohexyl residue is present in phosphonium salt (XXVIc). The reaction of the phosphonium salt XXVIa with cinnamaldehyde produced only 20% of the expected Schiff base.

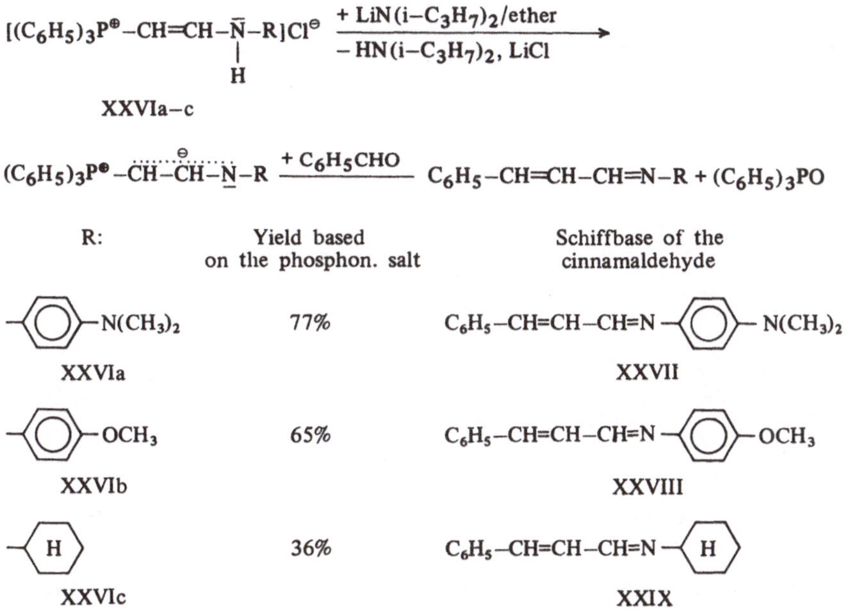

$$[(C_6H_5)_3P^{\oplus}-CH=CH-\underset{\underset{H}{|}}{\bar{N}}-R]Cl^{\ominus} \xrightarrow[-HN(i-C_3H_7)_2,\ LiCl]{+LiN(i-C_3H_7)_2/ether}$$

$$\text{XXVIa–c}$$

$$(C_6H_5)_3P^{\oplus}-\overset{\ominus}{\overset{\cdots\cdots}{CH}}-CH-\underset{}{\overset{}{N}}-R \xrightarrow{+C_6H_5CHO} C_6H_5-CH=CH-CH=N-R + (C_6H_5)_3PO$$

R:	Yield based on the phosphon. salt	Schiffbase of the cinnamaldehyde
⟨benzene⟩–N(CH₃)₂ XXVIa	77%	C₆H₅–CH=CH–CH=N–⟨benzene⟩–N(CH₃)₂ XXVII
⟨benzene⟩–OCH₃ XXVIb	65%	C₆H₅–CH=CH–CH=N–⟨benzene⟩–OCH₃ XXVIII
⟨cyclohexyl H⟩ XXVIc	36%	C₆H₅–CH=CH–CH=N–⟨cyclohexyl H⟩ XXIX

---

[a] The NMR-spectra of the phosphonium salts XXVIa–XXVIc indicate the enamine-structure and not the expected iminstructure $[(C_6H_5)_3P^{\oplus}-CH_2-CH=N-R]Cl^{\ominus}$.

Analogous relationships are found in the reactions of the phosphonium salts with ketones such as benzophenone. With the application of phosphonium salt XXVIa, 31% of the Schiff base XXX could be isolated, while with XXVIb only 4% of the desired product XXXI could be obtained. Finally, with β-carbonylmethylene phosphonium chloride, which contains an unprotected carbonyl group, no further reaction occurs.

$$[(C_6H_5)_3P^{\oplus}-CH{=}CH-N-R]Cl^{\ominus} \xrightarrow[- HN(i-C_3H_7)_2, \ LiCl]{+ LiN(i-C_3H_7)_2/ether}$$

$$\underset{H}{|}$$

XXVIa–XXVIb

$$(C_6H_5)_3P^{\oplus}-\overset{\cdots\overset{\ominus}{\cdots}}{CH{-}CH}-N-R \xrightarrow{+ (C_6H_5)_2CO}$$

$$(C_6H_5)_3PO + (C_6H_5)_2C{=}CH-CH{=}N-R$$

R :	Yield based on the phosphon. salt	Final product

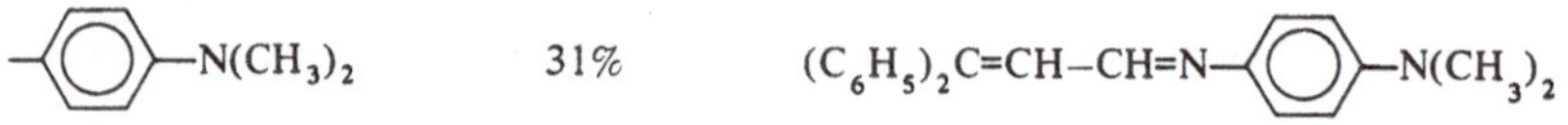

R: —⟨C₆H₄⟩—N(CH₃)₂   XXVIa   31%   Final product: $(C_6H_5)_2C{=}CH-CH{=}N-$⟨C₆H₄⟩$-N(CH_3)_2$ XXX

R: —⟨C₆H₄⟩—OCH₃   XXVIb   4%   Final product: $(C_6H_5)_2C{=}CH-CH{=}N-$⟨C₆H₄⟩$-OCH_3$ XXXI

$$(C_6H_5)_2P^{\oplus}-\overset{\cdots\overset{\ominus}{\cdots}}{CH{-}CH}-N-⟨C_6H_4⟩-N\overset{CH_3}{\underset{CH_3}{}} \ + \ \text{(fluorenone)}{=}O \xrightarrow{\text{ether}}$$

$$(C_6H_5)_3P{=}O \ + \ \text{(fluorenylidene)}{=}CH-CH{=}N-⟨C_6H_4⟩-N\overset{CH_3}{\underset{CH_3}{}}$$

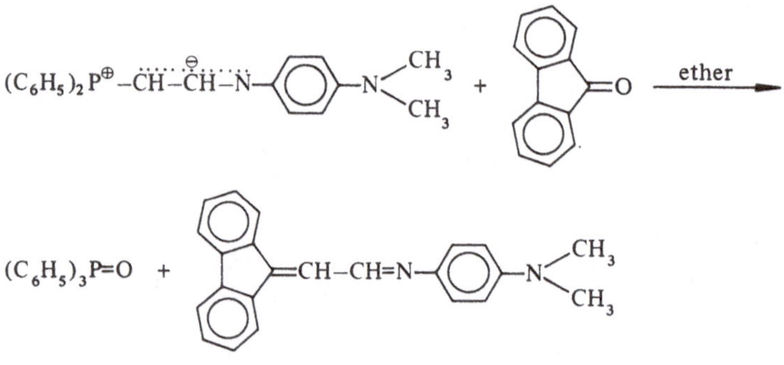

XXXII

When fluorenone is olefinated by the phosphorane produced from XXVIa, the corresponding Schiff base of the fluorenylidene acetaldehyde XXXII is isolated in 30% yield.

If the reaction is carried out in absolute ethanol with sodium ethoxide as the proton acceptor[13], the desired olefination products cannot be obtained from ketones

$$[(C_6H_5)_3P^{\oplus}-CH=CH-\overset{\underset{\textstyle H}{|}}{N}-C_6H_4-N\overset{\textstyle CH_3}{\underset{\textstyle CH_3}{<}}]\,Cl^{\ominus} \;+\; \text{(fluorenone)}\;=\!O \quad \xrightarrow[-C_6H_6,\ NaCl]{+NaOH/CH_2Cl_2}$$

XXVIa

$$(C_6H_5)_3PO \;+\; \text{(fluorenylidene)}\!=\!CH-CH=N-C_6H_4-N\overset{\textstyle CH_3}{\underset{\textstyle CH_3}{<}} \;+$$

XXXII

$$(C_6H_5)_2\overset{\underset{\textstyle O}{\|}}{P}-CH=CH-\overset{\underset{\textstyle H}{|}}{N}-C_6H_4-N\overset{\textstyle CH_3}{\underset{\textstyle CH_3}{<}} \;+\; (C_6H_5)_2\overset{\underset{\textstyle O}{\|}}{P}-CH_2-CH=N-C_6H_4-N\overset{\textstyle CH_3}{\underset{\textstyle CH_3}{<}}$$

XXXIIIa   XXXIIIb

such as benzophenone or cyclohexanone. The resulting phosphoranes were cleaved into triphenylphosphine oxide and the matching derivatives of the acetaldehyde.

The application of the phase transfer-catalyst method[14,15] led, in contrast to a decisive simplification of our olefination reaction.

In the two phase system, methylene chloride/water, the phosphonium salt XXVIa could be deprotonated with aqueous sodiumhydroxide and reacted with carbonyl compounds before its hydrolytic decomposition to the phosphine oxide dominated. The expected Schiff base could be obtained in 70% yield with benzaldehyde and 62% with cinnamaldehyde.

The reaction with fluorenone yielded, in addition to 23.4% of the phosphine oxide isomers XXXIIIa—XXXIIIb[b], 40% of the desired endproduct.

With this I come to the end of my presentation. It was not at all unusual, that in my research the original objective remained unattained simply because interesting byways led us in other directions. One could regard this as a "failed chemistry" but that would be exaggerated humility. If, nevertheless, something of value resulted, it was only because the opportunity was used as a path to other fruitful goals.

Finally, I wish to thank my coworkers who were involved with this set of problems. Mr. H. J. Schmidt, Mr. H. D. Frommeld, Mr. P. Suchanek, Mr. H. Reiff, Mrs. Hannelore Renner and Mrs. Ursula Schoch-Grübler. It was their skill and perseverance which made it possible to achieve new objectives.

---

[b] The NMR-spectrum shows the tautomeric forms to be present in the ratio of 80:20 (determined by the $(CH_3)_2N$-signals). An exact assignment of all peaks for both tautomers has not yet been possible. Because of phosphoruscoupling and possible cis-trans isomerization the spectrum is difficult to interpret.

G. Wittig

5. References

[1] Wittig, G., Schmidt, H. J., and Renner, H.: Chem. Ber. *95*, 2377 (1962).
[2] Wittig, G., and Frommeld, H. D.: Chem. Ber. *97*, 3541 (1964).
[3] Wittig, G., Frommeld, H. D., and Suchanek, P.: Angew. Chem. *75*, 978 (1963); Angew. Chem. internat. Edit. *2*, 683 (1963).
[4] Ziegler, K., Eberle, H., and Ohlinger, H.: Liebigs Ann. Chem. *504*, 94 (1933).
[5] Wittig, G., and Frommeld, H. D.: Chem. Ber. *97*, 3548 (1964).
[6] Reiff, H.: Dissertation Univ. Heidelberg 1966.
[7] Wittig, G., and Suchanek, P.: Tetrahedron Suppl. *8*, 347 (1966).
[8] Büchi, G., and Wüest, H.: Helv. Chim. Acta *50*, 2440 (1967).
[9] Wittig, G., and Reiff, H.: Angew. Chem. *80*, 15 (1968); Angew. Chem. internat. Edit. *7*, 7 (1968).
[10] Trippett, S., and Walker, D. M.: J. chem. Soc. (London) *1961*, 1266.
[11] Nagata, W., and Hayase, Y.: J. Chem. Soc. (London) *1969*, 460.
[12] Wittig, G., and Schoch-Grübler, U.: Publication in preparation.
[13] Wittig, G., and Haag, W.: Chem. Ber. *88*, 1654 (1955).
[14] Dejmlow, E. V.: Angew. Chem. *86*, 187 (1974).
[15] Märkl, G., and Merz, A.: Synthesis *1973*, 295.

Received January 1, 1976

Dihetero-tricyclodecanes

2,6-Dihetero-adamantanes, 2,7-Dihetero-isotwistanes, 2,7-Dihetero-twistanes,
2,8-Dihetero-homotwistbrendanes and 2,6-Dihetero-tricyclo[3.3.2.0^{3,7}]decanes

Dr. Camille Ganter

Laboratorium für Organische Chemie der Eidgenössischen Technischen Hochschule Zürich,
Universitätstraße 16, CH-8092 Zürich.

Contents

1. Introduction

Restriction among the great number of possible dihetero-tricyclodecanes to such with a carbocyclic 8-membered ring (cyclooctane) as basic skeleton, which is crosswise bridged by two heteroatoms, and restriction to 5-, 6- and 7-membered heterocyclic rings, isomeric dihetero-tricyclodecanes of the following five different structural types are possible: 2,6-dihetero-adamantane $(G\ 1^{a)})^{b)}$, 2,7-dihetero-isotwistane $(G\ 2)^{c)}$, 2,7-dihetero-twistane $(G\ 3)^{d)}$, 2,8-dihetero-homotwistbrendane $(G\ 4)^{e)}$ and 2,6-dihetero-tricyclo[3.3.2.0^{3,7}]decane $(G\ 5)^{f)}$.

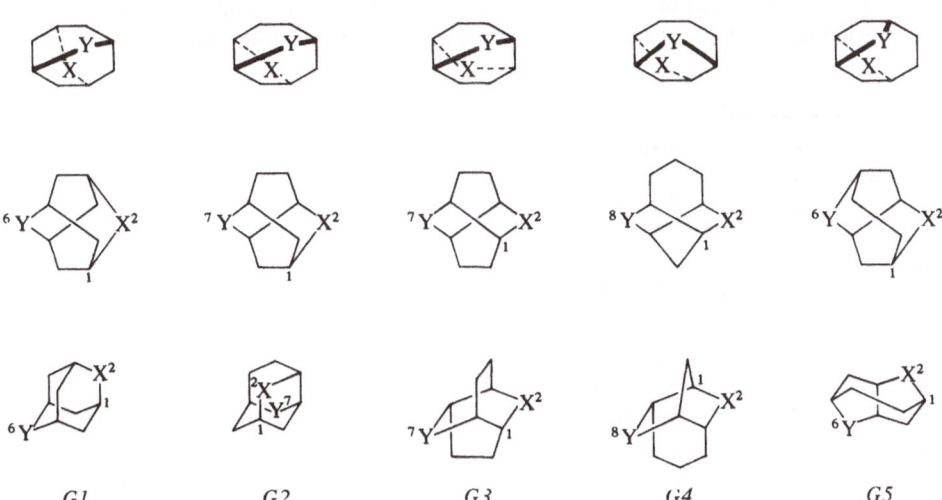

G1	G2	G3	G4	G5

This review summarizes the results on dihetero-tricyclodecanes of the types $G\ 1 - G\ 5$ known at present with particular emphasis on the synthesis of compounds from suitable 9-heterobicyclo[3.3.1]- and 9-heterobicyclo[4.2.1]nonanes as well as by molecular rearrangements of dihetero-tricyclodecanes involving neighboring group participation. Furthermore spectral data are discussed mainly in connection with structural assignments.

1.1. 2,6-Dihetero-adamantanes[b)]

Hexamethylenetetramine (urotropine) was not only the first but also for many years also the sole compound whose adamantane structure (tricyclo[3.3.1.1^{3,7}]decane) was a certainty. Interestingly, hexamethylenetetramine was also the first organic compound whose structure was determined by X-Ray in 1923[8,9)].

Pure carbocyclic adamantane [$G\ 1$: X(2) = CH$_2$, Y(6) = CH$_2$][g)] was first isolated in 1933 by Landa and Macháček[15,16)] from a high boiling petroleum fraction. Its structure was proved by a sophisticated, although laborious synthesis by Prelog and Seiwerth[17)] in 1941. In spite of some improvements of the original procedure the synthesis of larger amounts of adamantane remained difficult. In a first attractive alternative Schleyer[18] obtained among others 12% of adamantane by aluminium-bromide-catalyzed transformation of endo-trimethylene norbornane, the hydrogenation product from dimeric cyclopentadiene. Under modified reaction conditions the yields of this and analogous hydrocarbon rearrangements could even be successfully improved[g].

Adamantanes in which one or several of the methylene groups or one or several of the bridgehead carbonatoms are replaced by heteroatoms have been (with the exception of hexamethylenetetramine) subjects of intensive interest for about 20 years[h] only.

The first synthesis of a 2,6-dihetero-adamantane [$G\ 1$: X(2) = O, Y(6) = O] was reported by Stetter and Meissner in 1966[20,21]. Up to now the following 2,6-dihetero-adamantanes ($G\ 1$) and/or derivatives thereof became known:

Y(6)	X(2)	Refs.
O	O	20–27)
S	O	21, 28)
Se	O	27, 29)
NR	O	30–36)
PR	O	37)
S	S	38)
Se	S	39)
NR	S	31, 32, 34, 40)
NR^2	NR^1	32, 41–44)

1.2. 2,7-Dihetero-isotwistanes[c]

The first synthesis of a tricyclo[4.3.1.0^{3,8}]decane, namely of pure carbocyclic isotwistane [$G\ 2$: X(2) = CH$_2$, Y(7) = CH$_2$], was reported by Vogt[1] in 1968. Almost at the same time 2,7-dihetero-isotwistanes ($G\ 2$) and/or derivatives thereof became available. Today a variety of such compounds are known:

Y(7)	X(2)	Ref.
O	O	24,25, 45–52)
S	O	46, 47, 53, 54)
NR	O	35, 36, 55)
O	S	27, 53, 56)
S	S	27, 47, 57)
NR	S	58)
O	NR	58)
S	NR	58)

1.3. 2,7-Dihetero-twistanes[d]

The first tricyclo[4.4.0.0^{3,8}]decane known was pure carbocyclic twistane [G 3: X(2) = CH$_2$, Y(7) = CH$_2$] synthesized by Whitlock, Jr. in 1962[2, 59]. The 2,7-dihetero-twistanes (G 3) and/or derivatives thereof so far known are:

Y(7)	X(2)	Ref.
O	O	24, 25, 45—47, 52, 60, 61)
S	O	27, 46, 47, 53, 54)
NR	O	35, 55, 58)
NR	S	58)

1.4. 2,8-Dihetero-homotwistbrendanes[e]

2,8-Dioxa-homotwistbrendane [G 4: X(2) = O, Y(8) = O] and some derivatives thereof[50] represent the only tricyclo[5.3.0.0^{3,9}]decanes known up today.

1.5. 2,6-Dihetero-tricyclo[3.3.2.0^{3,7}] decanes[f]

Neither carbocyclic [G 5: X(2) = CH$_2$, Y(6) = CH$_2$] nor heterocyclic compounds (G 5) of this structural type have been synthesized so far.

1.6. General Synthetic Aspects

In principle two different synthetic approaches may lead to dihetero-tricyclodecanes of the types G 1--G 5: starting either from suitably functionalized 9-heterobicyclo-[3.3.1]- or 9-heterobicyclo[4.2.1]nonanes by bridging with a second heteroatom (see 2.) or already from dihetero-tricyclodecanes G 1 – G 5 and derivatives thereof, respectively, by molecular rearrangements under neighboring group participation of the heteroatoms X or Y (see 3.). Subsequently further derivatives can be prepared (see 4.) from compounds obtained by either of the two general ways.

2. Syntheses: 9-Heterobicyclo[3.3.1]nonanes or 9-Heterobicyclo[4.2.1]-nonanes → Dihetero-tricyclodecanes

2.1. 2,6-Dihetero-adamantanes

2.1.1. Introduction

Of the many possible general pathways the following five A—E have already been successfully applied to the synthesis of 2,6-dihetero-adamantanes. They use either

21

saturated 9-heterobicyclo[3.3.1]nonanes *G 7* suitably substituted at C(3) or 9-heterobicyclo[3.3.1.]nona-2,6-dienes *G 8* as well as corresponding monoepoxides *G 9* and diepoxides *G 10* and *G 11* as starting materials.

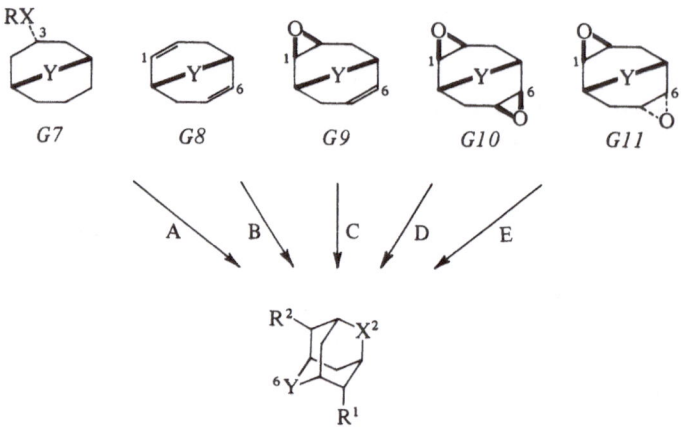

pathway A: *G 7* → *G 1* (R^1 = H, R^2 = H)
pathway B: *G 8* → *G 12* (R^1 ≠ H, R^2 ≠ H)
pathway C: *G 9* → *G 13* (R^1 = OH, R^2 ≠ H)
pathway D: *G 10* ↘ *G 14* (R^1 = OH, R^2 = OH)
pathway E: *G 11* ↗

2.1.2. 2,6-Dioxa-adamantanes

2.1.2.1. Pathway B (G 8 → G 12)[20–24, 26]

By the approach B Stetter and coworkers[20,21] successfully synthesized in 1966 the first 2,6-dihetero-adamantane, the 2,6-dioxa-adamantane (*14*). Oxymercuration of 9-oxabicyclo[3.3.1]nona-2,6-diene(*1*)[20–23, 48, 62, 63] with mercuric acetate in water yielded a cristalline acetoxymercuri compound, which after treatment with potassium iodide gave the corresponding iodomercuri compound. Subsequent iododemercuration led to a 4,8-diiodo-2,6-dioxa-adamantane, which by catalytic hydrogenation yielded unsubstituted 2,6-dioxa-adamantane (*14*). The authors did not discuss the configurations of the mercuri substituents. However, nmr.-data (see 5.2.3.1.) give conclusive evidence that in the compounds *2–4* both mercuri substituents are orientated towards one and the same O-atom O(2) or O(6)[24,26]. Further investigations[24,26] also proved that treatment of the acetoxymercuri- (*2*) or the therefrom obtained chloromercuri compound (*4*) with sodium borohydride (NaBH₄) in aqu. NaOH-solution in both cases yielded unsubstituted 2,6-dioxa-adamantane (*14*) and that iododemercuration of *3* gave not only one but both possible isomers *5* and *6*. The diiodide *5* as sole product was obtained by reaction of 9-oxabicyclo[3.3.1]-nona-2,6-diene (*1*) with mercuric oxide and iodine. Treatment of the diiodides

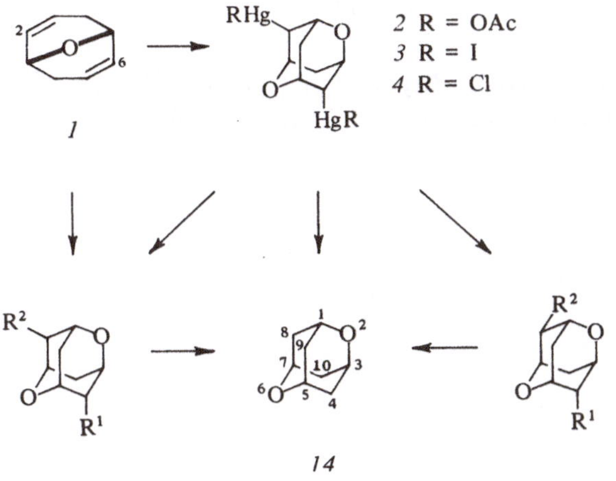

$$2 \quad R = OAc$$
$$3 \quad R = I$$
$$4 \quad R = Cl$$

	R^1	R^2			R^1	R^2
5	I	I		6	I	I
7	OAc	I		8	OAc	I
9	OAc	OAc		10	OAc	OAc
11	Br	Br				
12	Cl	Cl				
13	OH	OH				

5 and 6 with silver acetate in acetic acid yielded, depending on the reaction time, among others the monoacetates 7 and 8 or the diacetates 9 and 10[25) i)] (see 3.2.2.2.).

Further syn-4,8-disubstituted 2,6-dioxa-adamantanes were prepared from diene 1 applying pathway B by Cuthbertson and MacNicol[22)]: → dibromide 11 → 14 and by Averina et al.[23)]: → dibromide 11, dichloride 12 and diol 13. The latter 13 was also obtained by the pathways D and E (see 2.1.2.3.).

2.1.2.2. Pathway C (G 9 → G 13)[27)]

Treatment of exo,exo-2,3-epoxy-9-oxabicyclo[3.3.1]non-6-ene (15)[36)] with mercuric acetate yielded the 2,6-dioxa-adamantane derivative 16, which was directly demer-

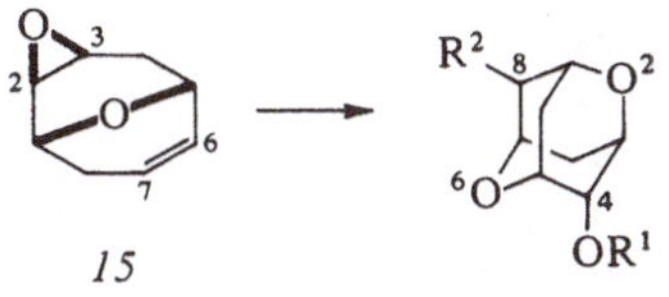

	R^1	R^2
16	H	HgOAc
17	H	H
18	Ac	H
19	Ts	H

curated with NaBH$_4$ to the adamantanol 17. The latter was also characterized as its acetate 18 and tosylate 19. LiAlH$_4$-reduction of 19 gave again the alcohol 17.

23

2.1.2.3. Pathways D (G 10 → G 14) and E (G 11 → G 14) [27]

Refluxing a solution of exo,exo-2,3-exo,exo-6,7-diepoxide *20* (syn-exo)[36] in 0.5N HCl resultated in the formation of adamantane-diol *13* as sole product, *i.e.* in

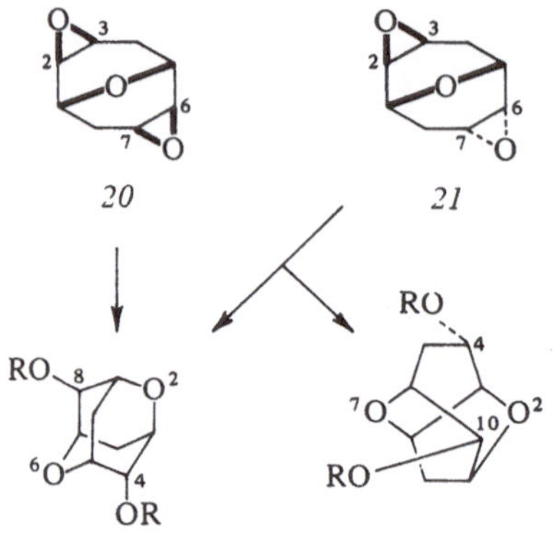

13 R = H
 9 R = Ac

22 R = H
23 R = Ac

20 both epoxides were cleaved only by trans-opening of the oxygen-carbon bonds to C(3) and C(7), resp. However, analogous treatment of the exo,exo-2,3-endo,endo-6,7-diepoxide (anti) *21*[36] yielded not only *13* but also $4^{O(7)},10^{O(7)}$-dihydroxy-2,7-dioxa-isotwistane (*22*)[l]. The endo-epoxide ring in *21* was therefore opened in two ways: cleavage of the bond O—C(6) → endo-OH at C(7) and cleavage of the bond O—C(7) → endo-OH at C(6), each by trans-opening of the epoxide ring. Both diols *13* and *22* were also characterized as their diacetates $9^{k)}$ and $23^{k)}$.

2.1.3. 2-Oxa-6-thia-adamantanes: Pathway B (G 8 → G 12) [21, 28]

By transannular addition of sulfur dichloride to 9-oxabicyclo[3.3.1]nonan-2,6-diene (*1*), $4^{O(2)}, 8^{O(2)}$-dichloro-2-oxa-6-thia-adamantane (*24*)[l] was obtained in approx. 60% yield, which by LiAlH$_4$-reduction was easily converted to unsubstituted 2-oxa-

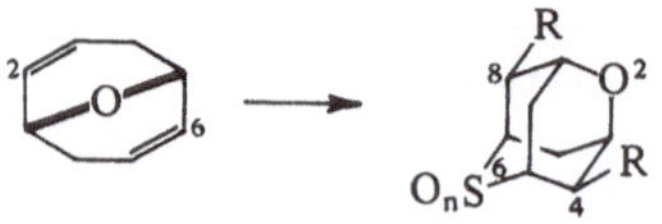

	n	R
24	0	Cl
25	0	OH
26	0	OAc
27	0	OCH$_3$
28	0	H
29	1	OH
30	2	Cl
31	2	OH
32	2	OAc
33	2	H

6-thia-adamantane (28)[21,28]. On the other hand, the dichloride 24 could be transformed almost quantitatively into the $4^{O(2)}$, $8^{O(2)}$-diol 25[1], which consequently was acetylated to the diacetate 26[28] and methylated to the dimethoxy compound 27[21]. Oxidation of diol 25 with hydrogen peroxide in acetic acid led depending on the reaction conditions either to sulfoxide 29 or directly to the corresponding sulfone 31[28] from which also the diacetate 32 was prepared. The sulfones 30 (R = Cl) and 33 (R = H) were obtained by analogous oxidation of the corresponding sulfides 24 and 28[21].

2.1.4. 2-Oxa-6-selena-adamantanes: Pathway B (G 8 → G 12)[27, 29]

Analogous to conversions by Lautenschlaeger[39] (see 2.1.8.) treatment of 9-oxa-bicyclo[3.3.1]nona-2,6-diene (1) with selenium dichloride led to $4^{O(2)}$,$8^{O(2)}$-di- • chloro-2-oxa-6-selena-adamantane (34) in 85% yield, which by reaction with silver

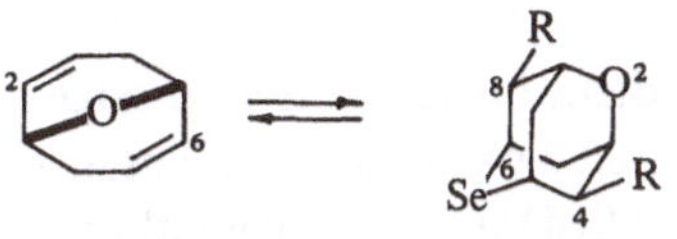

	R
34	Cl
35	OAc
36	OH
37	H

1

acetate in acetonitrile/acetic acid was converted almost quantitatively into the diacetate 35[m]. LiAlH4-reduction of 35 gave the diol 36. Attempted conversion of the dichloride 34 to unsubstituted 2-oxa-6-selena-adamantane (37) by LiAlH4-treatment failed, diene 1 was the only reaction product.

2.1.5. 2-Oxa-6-aza-adamantanes

2-Oxa-6-aza-adamantanes have been prepared by all five pathways A—E.

2.1.5.1. Pathway A (G 7 → G 1) [30, 33]

By treatment of N(9)-methyl-7-ethoxy-granat-3α-ol (38) with hydrogen bromide Stetter & Mehren[30] obtained in 1967 unsubstituted N(6)-methyl-2-oxa-6-aza-adamantane (40). This was the first synthesis of a 2-oxa-6-aza-adamantane.

38 R^1 = CH₃, R^2 = OC₂H₅
39 R^1 = COCH₃, R^2 = H

40 R = CH₃
41 R = COCH₃

Kashman and Benary[33] reported the successful ring closure of the bicyclic alcohol *39* by lead tetraacetate (without or with iodine added) to the N(6)-acetyl-2-oxa-6-aza-adamantane (*41*).

2.1.5.2. Pathway B (G 8 → G 12)[31, 32, 36]

Treatment of N(9)-formyl-9-azabicyclo[3.3.1]nona-2,6-diene (*42*) with bromine in methanolic KOH-solution gave N(6)-formyl-4$^{N(6)}$,8$^{N(6)}$-dibromo-2-oxa-6-aza-adamantane (*43*) in 56% yield. Subsequent LiAlH$_4$-reduction gave 81% of unsubstituted N(6)-methyl-2-oxa-6-aza-adamantane (*40*)[36].

	R^1	R^2
43	CHO	Br
40	CH$_3$	H

N(9)-Phenylsulfonyl-9-azabicyclo[3.3.1]nona-2,6-diene (*44*) is another suitable starting material for the synthesis of C(4),C(8)-disubstituted 2-oxa-6-aza-adamantanes, the two substituents being equal[36]. The configurations at C(4) and C(8) may either

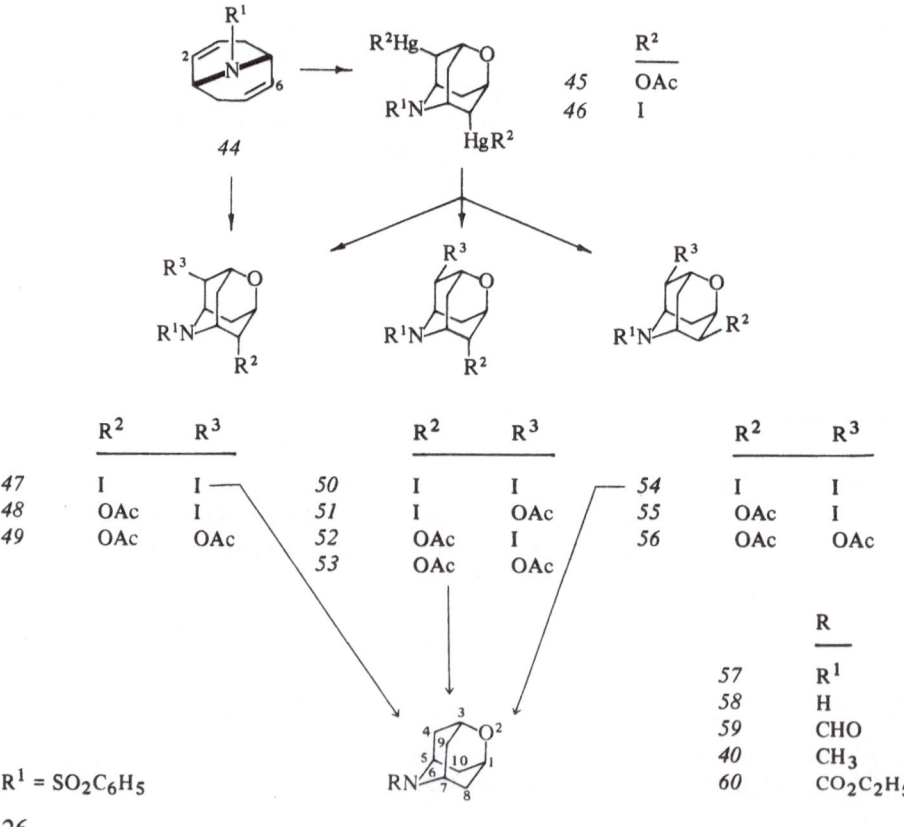

	R^2
45	OAc
46	I

	R^2	R^3
47	I	I
48	OAc	I
49	OAc	OAc

	R^2	R^3
50	I	I
51	I	OAc
52	OAc	I
53	OAc	OAc

	R^2	R^3
54	I	I
55	OAc	I
56	OAc	OAc

	R
57	R^1
58	H
59	CHO
40	CH$_3$
60	CO$_2$C$_2$H$_5$

R^1 = SO$_2$C$_6$H$_5$

be the same or differ. Oxymercuration of diene *44* by mercuric acetate in water/ methanol yielded almost quantitatively the tricyclic diacetoxymercuri compound *45*, which was easily converted to the corresponding diiodomercuri compound *46*. Iododemercuration of *46* gave a mixture of the three at C(4) and C(8) isomeric diiodo-2-oxa-6-aza-adamantanes *47* (10%), *50* (34%) and *54* (23%), which could be separated by fractional crystallization and chromatography on silicagel. Diiodide *47* as sole product was obtained in 79% yield by reaction of diene *44* with mercuric oxide and iodine. All three diiodides *47*, *50* and *54* were converted to unsubstituted N(6)- phenylsulfonyl-2-oxa-6-aza-adamantane (*57*) by reaction with desactivated raney-nickel C in methanolic KOH-solution (yields: 82–88%). Reduction of *57* with sodium in boiling isoamyl alcohol led to unsubstituted 2-oxa-6-aza-adamantane (*58*) (61%). By treatment of *58* with ethyl formate the formamide *59* (98%) was formed which by LiAlH$_4$-reduction gave N(6)-methyl-2-oxa-6-aza-adamantane (*40*). The corresponding urethane *60* (60%) was obtained by heating a solution of *40* and ethyl chloroformate in benzene.

Each of the three diiodides *47*, *50* and *54* was also treated with silver acetate in acetic acid[35)n)]. Starting from *47* one obtained after 230 hrs a product mixture which among others contained 28% monoiodo-acetate *48*[o)] but none of the corresponding diacetate *49*[o)]. Analogous treatment of *50*[n)] at 70° gave after 4 hrs among others 77% of the monoiodo-acetate *51*, after 89 hrs among others 50% of *51* and 20% of the diacetate *53* but no monoiodo-acetate *52*[o)]. Finally reaction of *54*[n)] for 9 hrs at 60° yielded among others monoiodo-acetate *55* (20% relative to reacted material) and diacetate *56* (66% relative to reacted material).

Treatment of the diacetoxymercuri compound *45* with NaBH$_4$ in aqu. NaOH-solution[36)] gave only 23% of unsubstituted N(6)-phenylsulfonyl-2-oxa-6-aza-ada-

	R^1	R^2	R^3
63	SO$_2$C$_6$H$_4$CH$_3$	Br	H
64	SO$_2$C$_6$H$_4$CH$_3$	H	Br
65	SO$_2$C$_6$H$_4$CH$_3$	H	H
58	H	H	H
58 · HCl	H · HCl	H	H
66	COC$_6$H$_5$	H	H

27

mantane (*57*) but at the same time 18% diene *44* and 21% bicyclic alcohol *61*. Diene *44* as sole product (65%) was obtained by reduction of *45* with raney-nickel C[36].

Ring closures to 2-oxa-6-aza-adamantanes starting from bicyclic dienes were also accomplished[31,32] by hydroxybromination of the 9-aza-diene *62* with N-bromosuccinimide (→ *63*: 53%) and by reaction of the 9-oxa-diene *1* with N,N-dibromo-p-tolylsulfonamide (→ *64*: 25%). Both dibromides were converted to *65* (raney-nickel/H_2), which itself by treatment with sodium in liquid ammonia, yielded unsubstituted 2-oxa-6-aza-adamantane (*58*), also characterized as its hydrochloride *58 · HCl* and N-benzoyl-derivative *66*.

2.1.5.3. Pathway C (G 9 → G 13) [36]

N(9)-Phenylsulfonyl-exo,exo-2,3-epoxy-9-azabicyclo[3.3.1]non-6-ene (*67*) is one of the suitable starting materials for the preparation of disubstituted [the substit-

	R^2	R^3	R^4
68	OH	HgOAc	H
69	OH	HgI	H
70	OH	I	H
71	OH	H	I
72	OH	H	H
48	OAc	I	H
52	OAc	H	I
73	OAc	H	H

$R^1 = SO_2C_6H_5$

67

uents at C(4) and C(8) being different] as well as monosubstituted [at C(4)] 2-oxa-6-aza-adamantanes. Oxymercuration of diene *67* with mercuric acetate in water/ methanol (→ *68*) followed by treatment with potassium iodide gave almost quantitatively the iodomercuri compound *69*. Its iododemercuration yielded a mixture of the two at C(8) epimeric iodo-alcohols *70* and *71*, which was acetylated and separated by chromatography: 32% *48* and 46% *52*. Both compounds in the presence of desactivated raney-nickel C were directly transformed into the monosubstituted N(6)-phenylsulfonyl-4$^{N(6)}$-hydroxy-2-oxa-6-aza-adamantane (*72*) (58–62%). The same alcohol *72* was also obtained by treatment of the acetoxymercuri compound *68* with $NaBH_4$ in aqu. NaOH-solution. By acetylation *72* was converted to the acetate *73*.

2.1.5.4. Pathways D (G 10 → G 14) and E (G 11 → G 14) [36]

Heating a solution of the syn-exo-diepoxide *74* in 0.5N HCl yielded 61% of N(6)-phenylsulfonyl-4$^{N(6)}$,8$^{N(6)}$-dihydroxy-2-oxa-6-aza-adamantane (*76*) as the sole tricyclic compound, which was also characterized as its diacetate *49*. Analogous treatment of the anti-epoxide *75* led to a mixture of adamantane-diol *76* (48%) and the isomeric N(7)-phenylsulfonyl-4$^{N(7)}$,10$^{N(7)}$-dihydroxy-2-oxa-7-aza-isotwistane (*77*) (22%). Both diols were also converted to the corresponding diacetates *49*[p] and *78*[p].

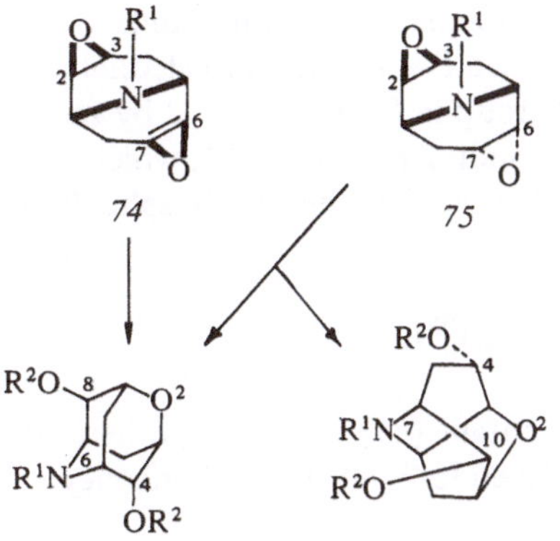

$$R^1 = SO_2C_6H_5$$

	R^2
76	H
49	Ac

	R^2
77	H
78	Ac

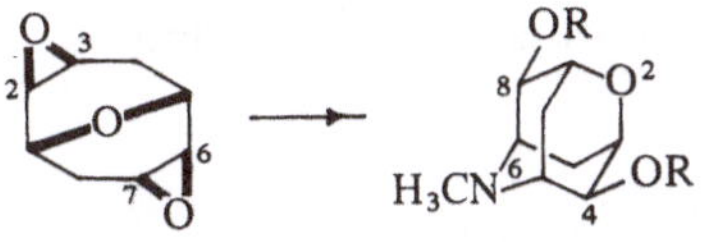

	R
79	H
80	Ac

Reaction of the syn-exo-diepoxide 20 (9-oxa) with methylamine under pressure gave 61% of N(6)-methyl-$4^{O(2)},8^{O(2)}$-dihydroxy-2-oxa-6-aza-adamantane (79), which by subsequent acetylation yielded the diacetate 80.

2.1.6. 2-Oxa-6-phospha-adamantanes: Pathway A (G 7 → G 1)[37]

Refluxing a solution of the endo-3-hydroxy-9-phosphabicyclo[3.3.1.]nonane 81 in the presence of lead tetraacetate yielded a mixture of three compounds, 2-oxa-6-

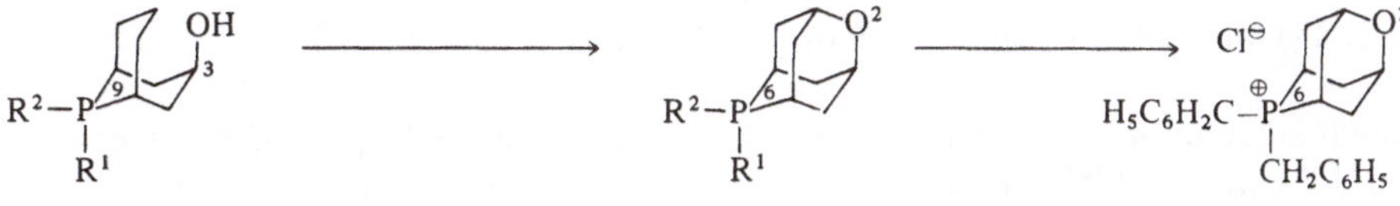

	R^1	R^2
81	=O	C_6H_{11}
82	=O	$CH_2C_6H_5$
83	$CH_2C_6H_5$	=O

	R^1	R^2
84	=O	C_6H_{11}
85	=O	$CH_2C_6H_5$
86	$CH_2C_6H_5$	:

29

phospha-adamantane *84* being the main product. Transannular reaction with lead tetraacetate was also performed on a mixture of the two P-epimeric endo-alcohols *82* and *83* leading to *85* as main product. The P=O group in *85* was reduced by tri-chlorosilane in benzene-solution to yield the corresponding phosphine *86*, which then on refluxing in acetonitrile with benzyl chloride gave the dibenzylphosphonium salt *87*.

2.1.7. 2,6-Dithia-adamantanes: Pathway B (*G 8 → G 12*)[38]

$4^{S(6)}, 8^{S(6)}, 9^{S(2)}, 10^{S(2)}$-Tetrachloro-2,6-dithia-adamantane (*89*) has been synthesized by transannular addition of sulfur dichloride to endo,endo-2,6-dichloro-9-thiabicy-

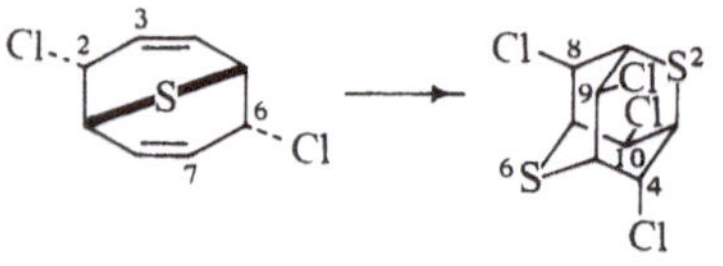

clo[3.3.1]nona-3,7-diene(*88*) (20–22%), which itself was obtained from cycloocta-tetraene by analogous reaction.

2.1.8. 2-Thia-6-selena-adamantanes: Pathway B (*G 8 → G 12*)[39]

The only compound with this skeleton, $4^{Se(6)}, 8^{Se(6)}, 9^{S(2)}, 10^{S(2)}$-tetrachloro-2-thia-6-selena-adamantane (*90*) was prepared by treatment of the diene *88*[38] with selenium monochloride (35%).

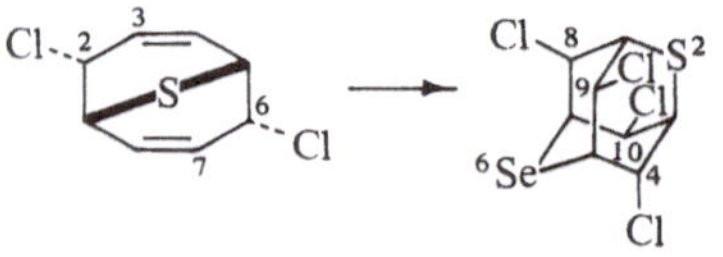

2.1.9. 2-Thia-6-aza-adamantanes: Pathway B (*G 8 → G 12*)[31, 32, 35, 40]

Transannular addition of sulfur dichloride to 9-azabicyclo[3.3.1]nona-2,6-dienes (*42*: N-formyl → *91*[40], *44*: N-phenylsulfonyl → *94*[40] and *62*: N-tolylsulfonyl → *96*[31, 32]) was applied as an entry to the 2-thia-6-aza-adamantane system. The chlorine atoms in *91* were easily exchangeable with sodium iodide in boiling diethyl ketone (→ *92*, 82%)[m] or with silver acetate in acetic acid (→ *93*, 70%)[m], resp. Under analogous reaction conditions the dichloride *94* was transformed into the diacetate *95* (95%)[m]. LiAlH$_4$-reduction of the diacetate *93* yielded 74% of N(6)-methyl-$4^{N(6)}, 8^{N(6)}$-dihydroxy-2-thia-6-aza-adamantane (*97*). The C(4) and C(8)

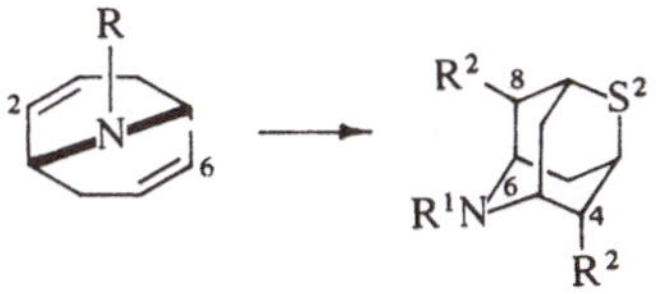

	R^1	R^2
91	CHO	Cl
92	CHO	I
93	CHO	OAc
94	$SO_2C_6H_5$	Cl
95	$SO_2C_6H_5$	OAc
96	$SO_2C_6H_4CH_3$	Cl
97	CH_3	OH
98	CH_3	H
99	$CO_2C_2H_5$	H
100	H	H

	R
42	CHO
44	$SO_2C_6H_5$
62	$SO_2C_6H_4CH_3$

unsubstituted 2-thia-6-aza-adamantane *98* was obtained by treatment of the dihalides *91* and *92* with $LiAlH_4$ (approx. 70% in each case). The tertiary amine *98* was easily transformed into its urethane *99* (85%), which by boiling in a HCl-solution gave the hydrochloride of 2-thia-6-aza-adamantane (*100 · HCl*). The free amine *100* was prepared by subsequent treatment with base.

2.1.10. 2,6-Diaza-adamantanes

2.1.10.1. Pathway A (G 7 → G 1)[41, 42]

A compound with 2,6-diaza-adamantane structure (diradical *106*) was described for the first time in 1969 by Rassat. The Hofmann-Löffler-Freytag reaction of the bi-

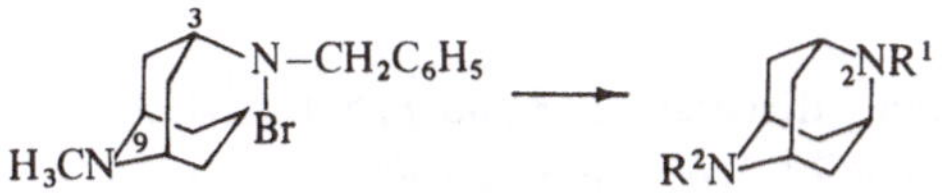

	R^1	R^2
102	CH_3	$CH_2C_6H_5$
103	CH_3	H
104	H	H
105	O	H
106	O	O

cyclic N-bromo-derivative *101* yielded the tricyclic compound *102*, which by hydrogenolysis gave N-monomethyl-2,6-diaza-adamantane (*103*). Subsequent treatment with potassium permanganate under basic conditions led to 2,6-diaza-adamantane (*104*), which on oxidation gave the monoradical *105* as well as the diradical *106*.

2.1.10.2. Pathway B (G 8 → G 12)[32, 43]

By pathway B, N(2),N(6)-ditolylsulfonyl-4$^{N(6)}$,8$^{N(6)}$-dibromo-2,6-diaza-adamantane (*107*) was synthesized (22%) by treatment of the bicyclic sulfonamide-diene *62* with N,N-dibromo-p-tolylsulfonamide in chloroform at low temperature. Hydrogenolysis

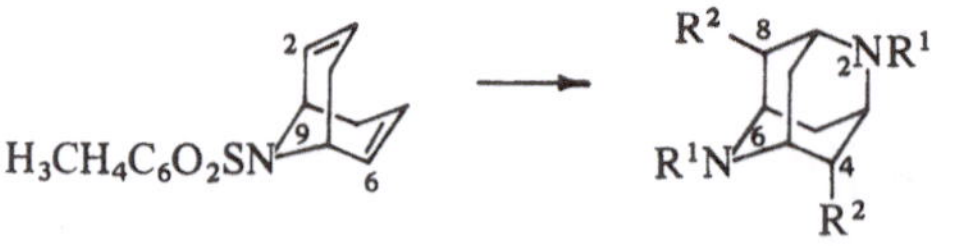

	R[1]	R[2]
107	$SO_2C_6H_4CH_3$	Br
108	$SO_2C_6H_4CH_3$	H
104	H	H
109	COC_6H_5	H
110	CH_3	H

of *107* yielded *108* (82%). The free 2,6-diaza-adamantane (*104*) was obtained from *108* by reaction with sodium in liquid ammonia (75%) and among others character-ized as its N(2),N(6)-dibenzoyl-derivative *109*. The N(2),N(6)-dimethyl-derivative *110* was prepared from *104* by the Leuckart-Wallach reaction (94%).

2.1.10.3. Pathway D (G 10 → G 14)[44]

Treatment of the syn-diepoxides *74* and *111* with methylamine under pressure yielded the $4^{N(6)}$, $8^{N(6)}$-dihydroxy-2,6-diaza-adamantanes *112* (94%) and *114* (63%), resp.,

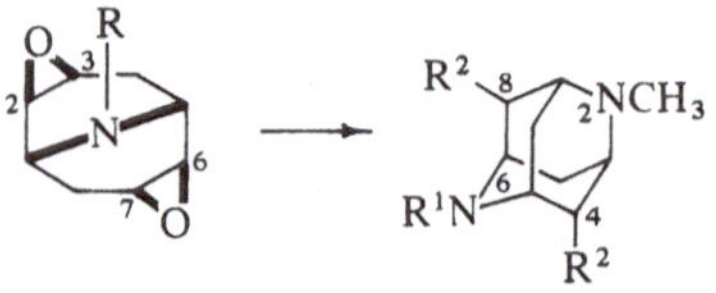

	R[1]	R[2]
112	$SO_2C_6H_5$	OH
113	$SO_2C_6H_5$	OAc
114	CHO	OH
115	CHO	OAc
116	CH_3	OH
117	CH_3	OAc
118	CH_3	Cl
110	CH_3	H

	R
74	$SO_2C_6H_5$
111	CHO

which each was converted to the corresponding diacetate *113* and *115*. LiAlH$_4$reduc-tion of the latter led to 2,6-diaza-adamantane-diol *116*, characterized also as its di-acetate *117*. The dichloro compound *118* (35%)[m] was obtained by reaction of the diol *116* with thionylchloride. *118* should easily be convertible to unsubstituted N(2), N(6)-dimethyl-2,6-diaza-adamantane (*110*).

2.2. 2,7-Dihetero-isotwistanes[c]

2.2.1. Introduction

Syntheses of various 2,7-dihetero-isotwistanes by bridging suitable bicyclononanes were successful applying one or several of the pathways F-M. Ring closures occurred either by nucleophilic attack at a double bond (in G 15, G 16 and G 17), at a car-bonyl carbon (in G 20) and at an epoxide ring (in G 21 and G 22) or by substitution of a leaving group (in G 18 and G 19).

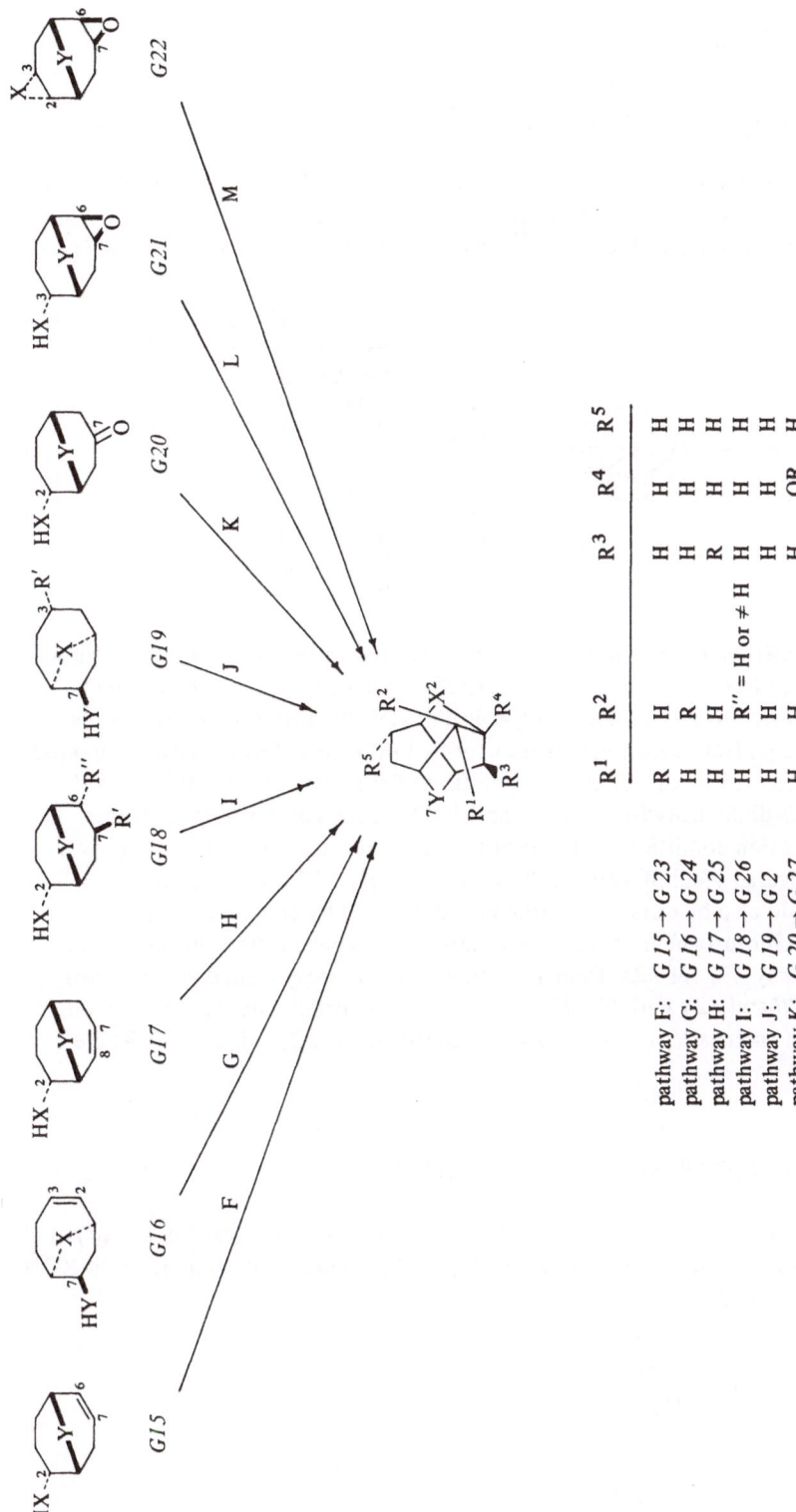

	R¹	R²	R³	R⁴	R⁵
	R	H	H	H	H
	H	R	H	H	H
	H	H	R	H	H
	H	R'' = H or ≠ H	H	H	H
	H	H	H	H	H
	OH	H	H	OR	H
	OH	H	H	H	H
	OH	H	H	H	OH

pathway F: G 15 → G 23
pathway G: G 16 → G 24
pathway H: G 17 → G 25
pathway I: G 18 → G 26
pathway J: G 19 → G 2
pathway K: G 20 → G 27
pathway L: G 21 → G 28
pathway M: G 22 → G 29

2.2.2. 2,7-Dioxa-isotwistanes

2.2.2.1. Pathway F (G 15 → G 23)[45, 46, 49–51]

endo-2-Hydroxy-9-oxabicyclo[3.3.1]non-6-ene (119)[48, 60, 64], easily accessible from cis,cis-cycloocta-1,5-diene, was the starting material chosen. Treatment of 119 with mercuric acetate in water yielded $10^{O(7)}$-acetoxymercuri-2,7-dioxa-isotwistane (120)[j],

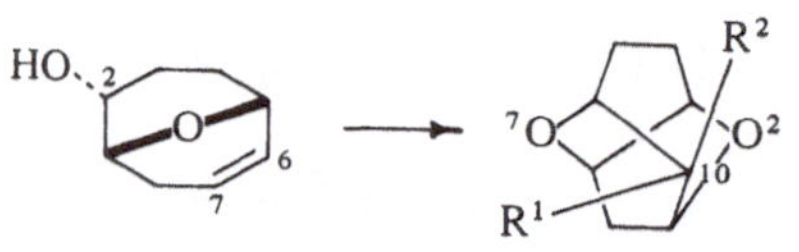

	R^1	R^2
120	HgOAc	H
121	HgBr	H
122	HgI	H
123	HgNO₃	H
124	HgCl	H
125	I	H
126	Br	H
127	H	I
128	H	H

which in aqu. KBr- or KI-solution, resp., was converted to the corresponding bromo-mercuri (121) and iodomercuri compound (122). By an analogous sequence diene 119 with mercuric nitrate in a 1% HNO_3-solution gave the nitratomercuri compound 123, which in aqu. NaCl-solution was transformed into the chloromercuri compound 124. Iododemercuration of 122 led to a mixture of the two epimeric $10^{O(7)}$- and $10^{O(2)}$-iodo-2,7-dioxa-isotwistanes (125 and 127), whose composition depended on the applied reaction conditions[q]. In all of the above reactions the yields were very high, e.g. the overall yield of 119 → 120 → 122 → 125 + 127 was 85%. The $10^{O(7)}$-iodide 125 could also be obtained as sole product (74%) by treatment of diene 119 with iodine in chloroform at room temperature. Analogous reaction in carbon tetrachloride gave only 47% of 125. Demercuration of the acetoxy- and chloromercuri compounds 120 and 124 with NaBH₄ as well as raney-nickel-reduction of the iodides 125 and 127 in each case gave in good yields unsubstituted 2,7-dioxa-isotwistane (128).

2.2.2.2. Pathway I [G 18 → G 26 (R" = H: G 2)][48]

2,7-Dioxa-isotwistane (128) was prepared by intramolecular substitution at C(7) of endo-2-hydroxy-exo-7-tosyloxy-9-oxabicyclo[3.3.1]nonane (129) in methanolic KOH-solution in 62% yield.

129   128

2.2.2.3. Pathway K (G 20 → G 27)[52]

Treatment of the tetrahydropyranylether *130* with methanesulfonic acid in methanol yielded beside 26% of the corresponding alcohol *131*, 45% of 1-methoxy-2,7-dioxa-

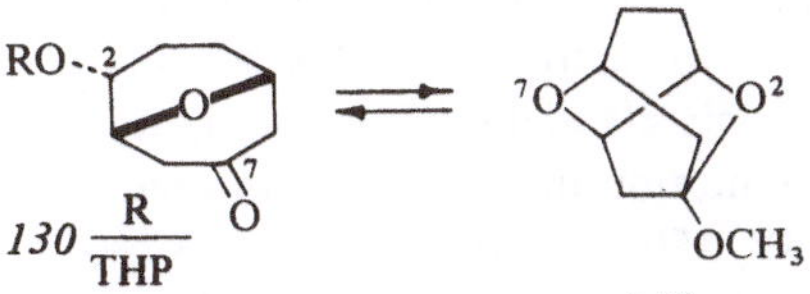

130 R̲ ̲
       THP
131 H
132 Ac

133

isotwistane (*133*), which by reaction in aqu. acid in a good yield gave the hydroxy-ketone *131*, which was also characterized as its acetate *132*.

2.2.2.4. Pathway L (G 21 → G 28)[48]

endo-2-Acetoxy-exo,exo-6,7-epoxy-9-oxabicyclo[3.3.1]nonane (*134*) when treated in a methanolic K_2CO_3-solution for 1 hr at 70° yielded a mixture of the correspond-

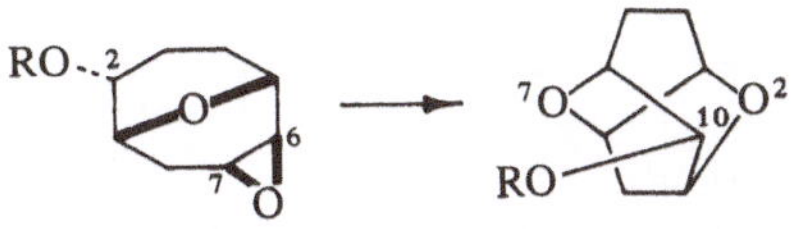

	R		R
134	Ac	*137*	H
135	H	*138*	THP
136	THP		

ing alcohol *135* and by intramolecular attack of the OH-group at the epoxide ring (trans-opening) $10^{O(7)}$-hydroxy-2,7-dioxa-isotwistane (*137*), which was converted to a separable mixture of the tetrahydropyranylethers *136* and *138*. Under more drastic conditions (21 hrs in a boiling K_2CO_3-solution) the epoxy-acetate *134* was transformed into the isotwistanol *137* in an almost quantitative yield. Furthermore, the latter was also obtained in a 96% yield by treatment of the tetrahydropyranyloxy-epoxide *136* with methanesulfonic acid and under the same conditions from the tricyclic tetrahydropyranylether *138*.

2.2.2.5. Pathway M (G 22 → G 29)

As already mentioned in 2.1.2.3., boiling a solution of the anti-diepoxide *21* in 0.5N HCl gave a mixture of $4^{O(7)},10^{O(7)}$-dihydroxy-2,7-dioxa-isotwistane (*22*) and ada-mantane-diol *13*. Both diols were also characterized as their acetates *23*[k] and *9*[k], respectively.

2.2.3. 2-Oxa-7-thia-isotwistanes: Pathway F (G 15 → G 23)

2.2.3.1. Sulfides[46, 53)]

Endo-2-Hydroxy-9-thiabicyclo[3.3.1]non-6-ene (*139*), easily prepared from cis,cis-cycloocta-1,5-diene, is a suitable starting material for the synthesis of 2-oxa-7-thia-isotwistanes. Oxymercuration of *139* with mercuric nitrate yielded the nonisolated nitratomercuri compound *140*, which by $NaBH_4$-reduction in basic solution gave directly unsubstituted 2-oxa-7-thia-isotwistane (*148*, 45% relative to *139*)[r)]. As by-products the bicyclic alcohol *139* (approx. 30%) and a mercuri compound (approx. 20%), to which structure *149* may be assigned, were isolated. Treating a basic solution of the nitratomercuri compound *140* with either NaCl, KBr, or KJ gave in each case in high yields the corresponding halomercuri compounds *142*, *143* and *144*, resp. By addition of bromine in chloroform to *143*, $10^{O(2)}$-bromo-2-oxa-7-thia-isotwistane (*145*) was obtained as the sole product (70%). However, analogous treatment of the iodomercuri compound *144* with iodine gave a separable mixture of the C(10)-epi-meric isotwistane-iodides *146* and *147* (approx. 85%) in the ratio of 2 : 1. Only $10^{S(7)}$-iodo-isotwistane *147* was formed by alkoxyiodination of the unsaturated bicyclic alcohol *139*. Best results were obtained with a solution of *139* in iodine, t-BuOK and t-BuOH, where approx. 40% of *147* could be isolated. With iodine in chloroform

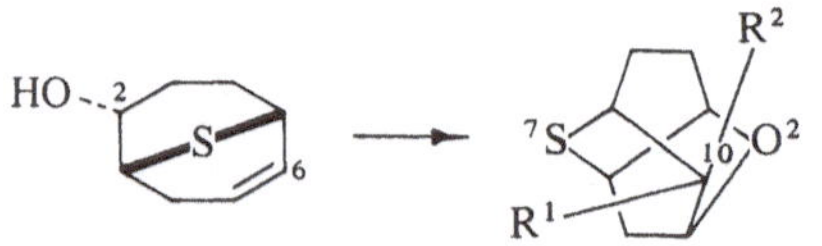

	R¹	R²
140	$HgNO_3$	H
141	HgOAc	H
142	HgCl	H
143	HgBr	H
144	HgI	H
145	H	Br
146	H	I
147	I	H
148	H	H

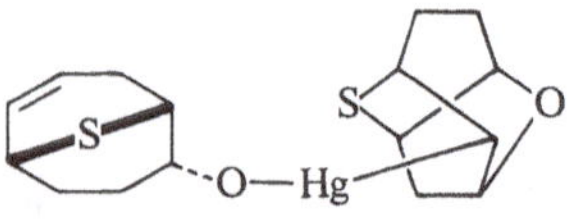

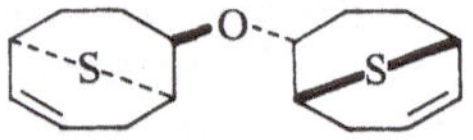

147 was formed only to about 12%, the main product being the ether *150* (32%). Both iodides *146* and *147* on $LiAlH_4$-reduction yielded unsubstituted 2-oxa-7-thia-isotwistane (*148*).

2.2.3.2. Sulfoxides and Sulfones[54)]

In addition to the above work on sulfides (see 2.2.3.1.) the behaviour of analogous sulfoxides and sulfons was also studied. The bicyclic sulfoxides *151* [the O-atom at S(9) orientated towards C(7)] and *152* [the O-atom at S(9) orientated towards C(3)] as well as the sulfone *153*, by reaction with mercuric nitrate and subsequent treatment with potassium iodide were converted to the corresponding tricyclic compounds: *151 → 154 → 155, 152 → 157 → 158* and *153 → 160 → 161*. Both sulfoxides *155* and *158* could be oxidized with hydrogen peroxide to the sulfone *161*. Sulfoxide *155* was also obtained as the sole product from $10^{S(7)}$-iodomercuri-2-oxa-7-thia-isotwistane (*144*) by oxidation with one equivalent of hydrogen peroxide.

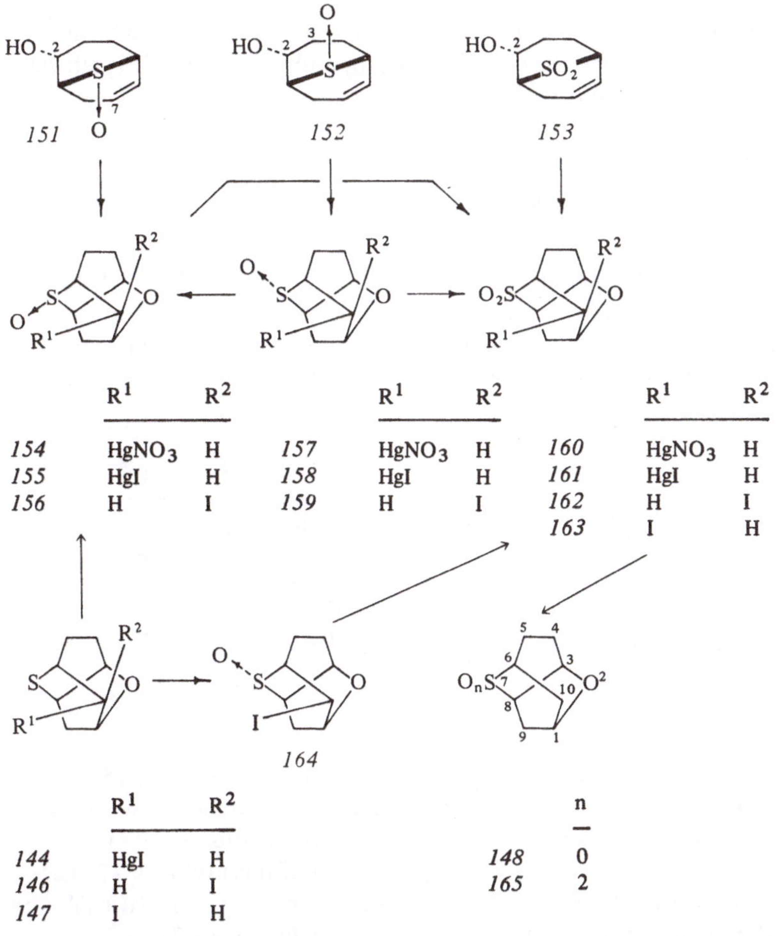

	R^1	R^2		R^1	R^2		R^1	R^2
154	HgNO$_3$	H	*157*	HgNO$_3$	H	*160*	HgNO$_3$	H
155	HgI	H	*158*	HgI	H	*161*	HgI	H
156	H	I	*159*	H	I	*162*	H	I
						163	I	H

	R^1	R^2		n
144	HgI	H	*148*	0
146	H	I	*165*	2
147	I	H		

Treatment of *155* with iodine in chloroform (iododemercuration) yielded the iodo-sulfoxide *156* as the sole product (approx. 40%), whereas, under analogous reaction conditions the S(7)-epimeric organomercuri compound *158* gave approx. 40% of the same iodide *156*. However, the S(7)-epimeric iodide *159* was also formed to a smaller extent (approx. 8%). Both sulfoxides *156* and *159* were converted to the same sulfone *162* by oxidation with hydrogen peroxide which by iododemercuration of *161* could also be obtained as an approx. 1 : 1-mixture with the $C(10)^{S(7)}$-epimer *163*. The latter compound was identical with the sulfone prepared via the sulfoxide

164 by oxidation of $10^{S(7)}$-iodo-2-oxa-7-thia-isotwistane (*147*) (see 2.2.3.1.). Reduction of the mixture of *162* and *163* with raney-nickel yielded the unsubstituted sulfone *165*, which was identical with the oxidation product of 2-oxa-7-thia-isotwistane (*148*).

2.2.4. 2-Oxa-7-aza-isotwistanes

Ring closures to tricyclic 2-oxa-7-aza-isotwistanes were carried out starting from the hydroxy-olefin *166* (pathway F), the corresponding hydroxy-epoxide *175* (pathway L) or the anti-diepoxide *75* (pathway M).

2.2.4.1. Pathway F (G 15 → G 23)[55]

Oxymercuration of N(9)-ethoxycarbonyl-endo-2-hydroxy-9-oxabicyclo[3.3.1]non-6-ene (*166*) with mercuric acetate in water/methanol yielded approx. 98% of the acetoxymercuri-isotwistane *167*, which by treatment with potassium iodide was easily transformed to the corresponding iodomercuri compound *168*. Iododemercuration

	R^1	R^2	R^3
167	$CO_2C_2H_5$	HgOAc	H
168	$CO_2C_2H_5$	HgI	H
169	$CO_2C_2H_5$	I	H
170	$CO_2C_2H_5$	H	I
171	$CO_2C_2H_5$	H	H
172	CH_3	H	H
173	H	H	H

(I_2/$CHCl_3$) of *168* led to a mixture of the two C(10)-epimeric iodides *169* (57%, $10^{N(7)}$-configuration) and *170* (37%, $10^{O(2)}$-configuration). Both iodides were reduced almost quantitatively with raney-nickel in methanolic KOH-solution to N(7)-ethoxy-carbonyl-2-oxa-7-aza-isotwistane (*171*). $LiAlH_4$-reduction of this urethane *171* gave the N(7)-methyl-derivative *172* (85%), whereas treatment of *171* in 20% HCl-solution for 4 days at reflux temperature finally yielded the hydrochloride of 2-oxa-7-aza-isotwistane (*173·HCl*, 96%). The free amine *173* was obtained by subsequent treatment with base.

2.2.4.2. Pathway L (G 21 → G 28)[55]

Acid-catalyzed intramolecular cyclization of the hydroxy-epoxide *175*, obtained by base-hydrolysis of the acetate *174*, yielded 97% of isotwistan-$10^{N(7)}$-ol *176*.

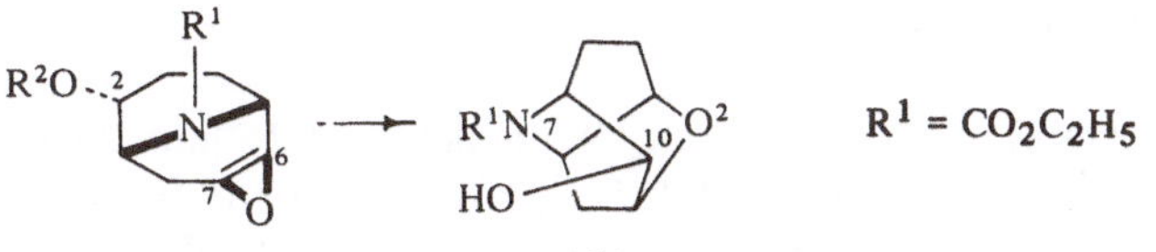

$R^1 = CO_2C_2H_5$

176

	R^2
174	Ac
175	H

2.2.4.3. Pathway M (G 22 → G 29)[36]

As already described in 2.1.5.4. a mixture of N(7)-phenylsulfonyl-4$^{N(7)}$,10$^{N(7)}$-dihydroxy-2-oxa-7-aza-isotwistane (*77*, 22%) and the isomeric adamantane-diol *76* (61%), both compounds also characterized as their acetates *78*[p] and *49*[p], was obtained by refluxing a solution of the anti-diepoxide *75* in 0.5N HCl.

2.2.5. 2-Thia-7-oxa-isotwistanes

2.2.5.1. Pathway F (G 15 → G 23)[27, 56]

Acid-catalyzed intramolecular addition of the free HS-group of the mercaptane *177* to the double bond C(6)-C(7) yielded almost quantitatively the tricyclic compound *179*. Treatment of it with HCl/ethanol gave the corresponding ketal *180*, with thionylchloride the two isomeric chlorides *181* (isotwistane-derivative) and *184* (bicyclo-[4.2.1]nonane-derivative). The latter was obviously formed by chlorination of the free hydroxy-ketone, which corresponds to the hemiketal *179*. Reduction of *181* by sodium in tetrahydrofuran finally yielded unsubstituted 2-thia-7-oxa-isotwistane (*182*), the very first compound with 2,7-dihetero-isotwistane structure[56].

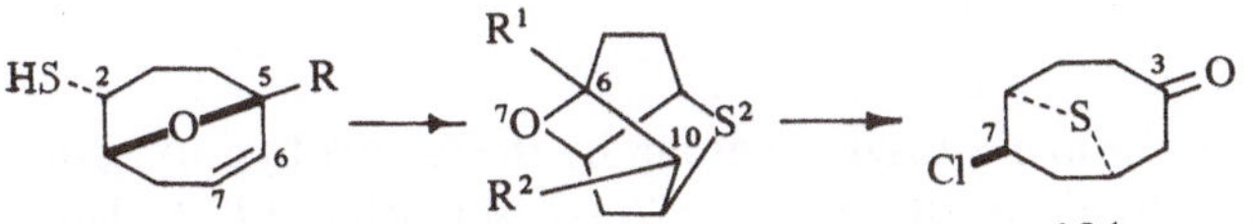

184

	R		R^1	R^2
177	OH	*179*	OH	H
178	H	*180*	OC$_2$H$_5$	H
		181	Cl	H
		182	H	H
		183	H	Br

Using an analogous starting material, the bicyclic mercaptane *178*[53] reacted with bromine in carbontetrachloride at −18° yielding approx. 95% of 10$^{O(7)}$-bromo-2-thia-7-oxa-isotwistane (*183*)[27].

2.2.5.2. Pathway G (G 16 → G 24)[27]

Oxymercuration (mercuric acetate) of endo-7-hydroxy-9-thiabicyclo[4.2.1]non-2-ene (188), a starting material prepared by the synthesis 185 → 186 → 187 → 188 from the known endo-7-acetoxy-9-thiabicyclo[4.2.1]nonan-3-one (185)[56] yielded the

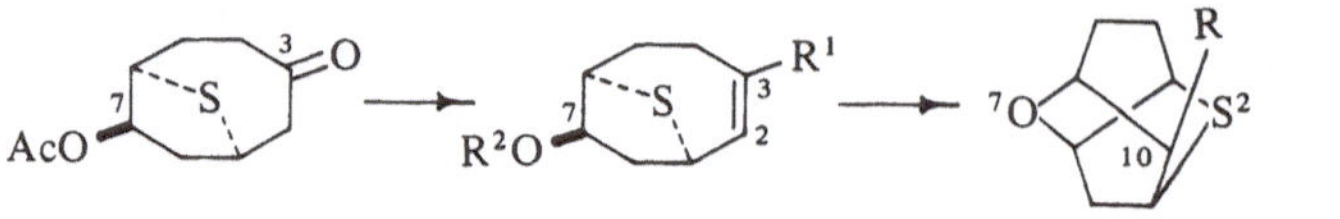

	R¹	R²		R
186	Ac	Ac	189	HgOAc
187	H	Ac	182	H
188	H	H		

acetoxymercuri compound 189, which by NaBH₄-reduction was transformed to unsubstituted 2-thia-7-oxa-isotwistane (182).

2.2.5.3. Pathway H (G 17 → G 25)[53]

A further derivative, the 9^{O(7)}-bromo-2-thia-7-oxa-isotwistane (191), was obtained by reaction of endo-2-mercapto-9-oxabicyclo[3.3.1]non-7-ene (190) with bromine.

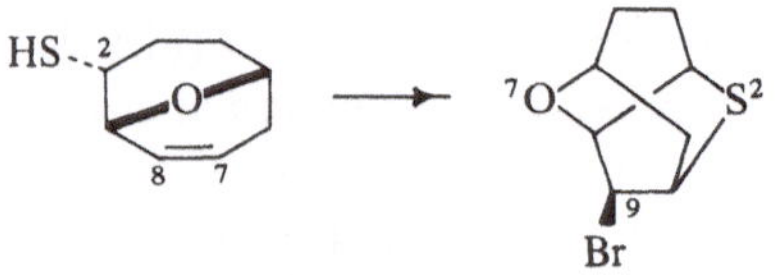

2.2.5.4. Pathway J (G 19 → G 2)[27]

Intramolecular substitution of the exo-3-tosyloxy group by the endo-7-hydroxy group in 195 opened a further access to unsubstituted 2-thia-7-oxa-isotwistane (182). The starting material 195 was obtained by NaBH₄-reduction of 185[56] (→ 192 + 193)

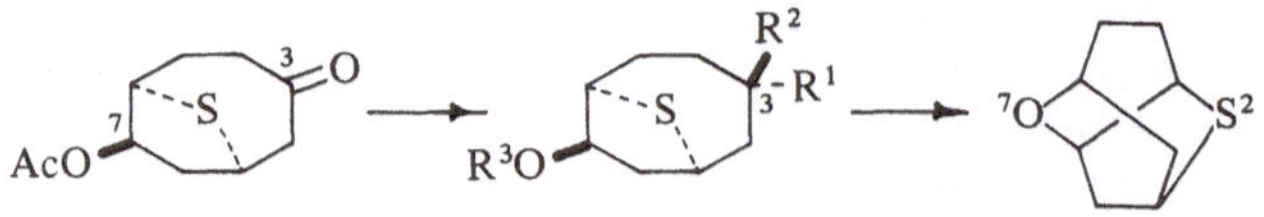

	R¹	R²	R³	182
192	OH	H	Ac	
193	H	OH	Ac	
194	OTs	H	Ac	
195	OTs	H	H	

followed by tosylation of the former (→ *194*). *195* formed by treatment with base was not isolated.

2.2.6. 2,7-Dithia-isotwistanes: Pathway F (*G 15 – G 23*)[57]

For the preparation of 2,7-dithia-isotwistanes the mercaptane *196* as well as the disulfide *197* were used as starting materials. Treatment of *196* with bromine or

	R			R
196	H		*198*	Br
			199	Cl
197			*200*	H

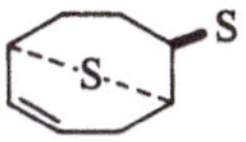

chlorine in carbontetrachloride in each case yielded approx. 90% $10^{S(7)}$-bromo(or chloro, resp.)-2,7-dithia-isotwistane (*198* or *199*, resp.). Analogous reaction of the disulfide *197* with chlorine also gave *199*, however, in a much lower yield. The unsubstituted 2,7-dithia-isotwistane (*200*) was obtained by $LiAlH_4$-reduction of the bromide *198* as well as of the chloride *199*.

2.2.7. 2-Aza-7-oxa-isotwistanes: Pathway F (*G 15 → G 23*)[58]

Entry to the 2-aza-7-oxa-isotwistane system was successful by olefin amination of the bicyclic amine *201* with mercuric acetate in dimethyl sulfoxide. The $10^{O(7)}$-acetoxy-

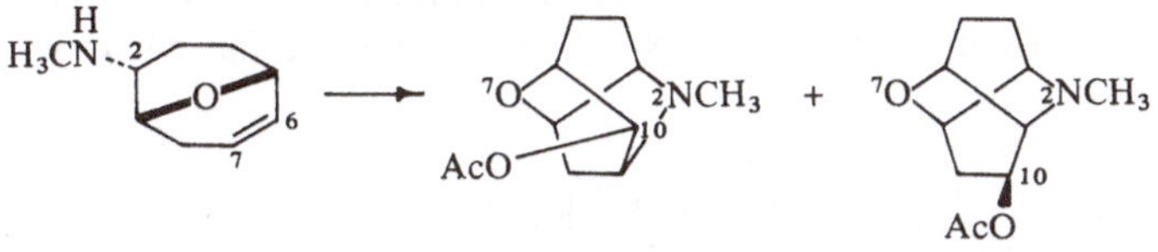

201	*202*	*203*

isotwistane *202* was obtained in a 72% yield beside traces of the corresponding twistane-acetate *203*.

2.2.8. 2-Aza-7-thia-isotwistanes

2.2.8.1. Pathway F (*G 15 → G 23*)[58]

Olefin amination of the bicyclic unsaturated secondary amine *204* with mercuric acetate in dimethyl sulfoxide yielded up to 85% of the $10^{S(7)}$-acetoxy-2-aza-7-thia-

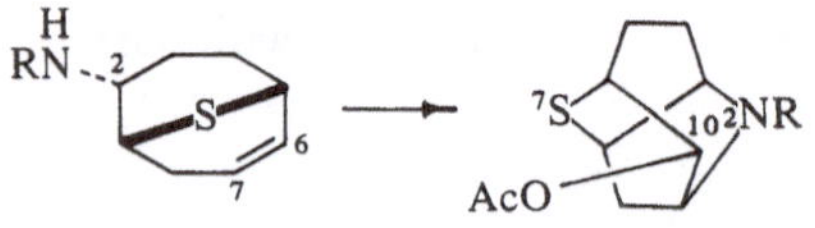

C. Ganter

	R			R
204	CH$_3$		*206*	CH$_3$
205	H		*207*	H

isotwistane *206*, whereas by analogous reaction of the bicyclic primary amine *205* the corresponding isotwistane *207* was formed only in a moderate yield.

2.2.8.2. Pathway I (G 18 → G 26)[58]

Cyclization of the bicyclic bromo-amine *208* in 1,2-dichloroethane in the presence of solid Na$_2$CO$_3$ yielded over 90% of the $10^{N(2)}$-bromo-isotwistane *209*.

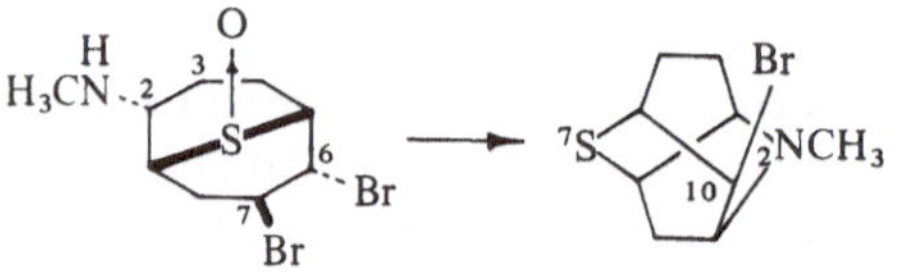

208	*209*

2.3. 2,7-Dihetero-twistanes

2.3.1. Introduction

For the synthesis of 2,7-dihetero-twistanes by bridging a suitable heterobicyclononane with a second heteroatom, only three pathways (N–P) were applied so far, whereby

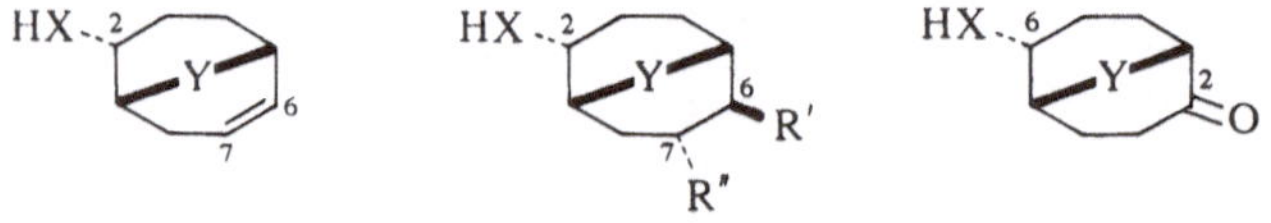

G15 *G30* *G31*

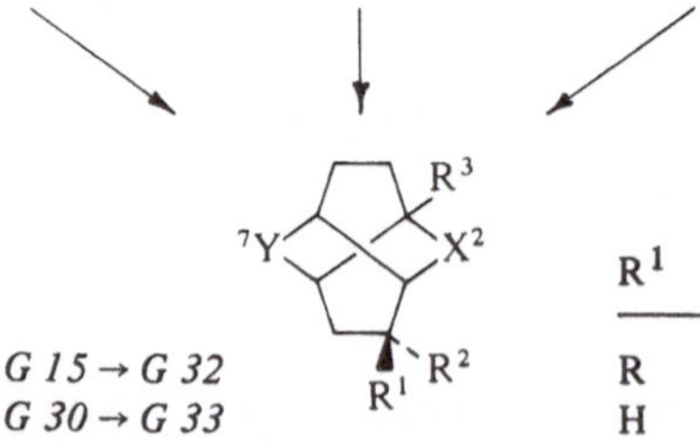

	R^1	R^2	R^3
pathway N: G 15 → G 32	R	H	H
pathway O: G 30 → G 33	H	R'' = H or ≠ H	H
pathway P: G 31 → G 34	H	H	OR

42

ring closures were performed either by nucleophilic attack at the double bond C(6)-C(7) (in *G 15*) and at a carbonyl carbon (in *G 31*) or by substitution of a leaving group (in *G 30*).

2.3.2. 2,7-Dioxa-twistanes

2.3.2.1. Pathway O (G 30 → G 33)[52, 60]

Treatment of endo-2-hydroxy-exo-6-iodo-9-oxabicyclo[3.3.1]nonane (*210*) in methanolic KOH-solution at reflux temperature yielded a mixture of 68% of the dehydrohalogenated compound *119* and 8% of 2,7-dioxa-twistane (*212*)[s], the product of intramolecular substitution. The ratio of elimination to substitution could lean strongly in favor of the latter by refluxing *210* in pyridine: 47% of the twistane *212* and only 28% of the unsaturated bicyclic alcohol *119* were formed[60].

	R¹	R²		R
210	I	H	*212*	H
211	OTs	CH₃	*213*	CH₃

By the same pathway reaction of endo-2-hydroxy-exo-6-tosyloxy-endo-7-methyl-9-oxabicyclo[3.3.1]nonane (*211*) in methanolic NaOH-solution, $4^{O(2)}$-methyl-2,7-dioxa-twistane[j] (*213*) could be obtained in high yield[52].

2.3.2.2. Pathway P (G 31 → G 34)[52]

On the basis of the results described in 2.2.2.3., namely, the easy intramolecular ketal formation of *131* to *133*, the question arose whether endo-6-hydroxy-9-oxabicyclo[3.3.1]nonan-2-one (*214*)[48] might analogously be converted to the corresponding l-methoxy-2,7-dioxa-twistane (*219*). The experiments, even under drastic acidic conditions in methanol (5.5N HCl in CH₃OH, 2 days at 80°), showed that no ring closure *214* → *219* occurred. Introduction of an endo-methyl group at C(7), however, strongly favors the formation of the tricyclic hemiketal with twistane structure. In the IR-spectrum (CHCl₃) of the alcohol obtained by base-treatment of endo-6-acetoxy-endo-7-methyl-9-oxabicyclo[3.3.1]nonan-2-one (*215*) only a weak carbonyl absorption (1712 cm⁻¹) can be observed. The IR-spectrum of the crystalline compound in KBr showed no carbonyl absorption at all. In the NMR-spectrum (CDCl₃) the signals of two secondary methyl groups at δ = 1.01 ppm and 1.16 ppm (ratio 1 : 5) are observed. All these results reflect the equilibrium between the bicyclic hydroxyketone *216* and the tricyclic hemiketal *218*. Treatment of *216* + *218* with methano-

43

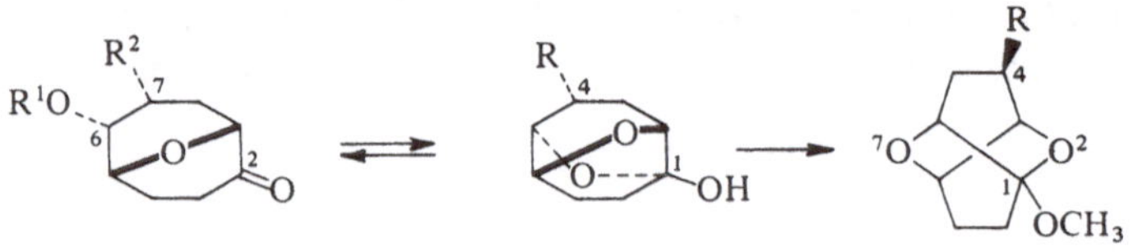

	R^1	R^2			R			R
214	H	H		217	H		219	H
215	Ac	CH$_3$		218	CH$_3$		220	CH$_3$
216	H	CH$_3$						

lic HCl-solution yielded 1-methoxy-4$^{O(2)}$-methyl-2,7-dioxa-twistane (*220*). Model studies on *216* show that in a chair-conformation the methyl group at C(7) would be axial. A boat- or twist-conformation of the six-membered ring, which therefore seems much more likely, is a necessary condition for the ring closure in the case of a twistane-derivative. As demonstrated by the above reactions ring closures starting from bicyclo[3.3.1]nonanes to twistanes [see *210* → *212* and *211* → *213* (2.3.2.1.), *214* → *217*, *216* → *218*] require more conformational changes than those involving iso-twistanes [see *129* → *128* (2.2.2.2.) and *131* → *133* (2.2.2.3.)].

The synthesis of the twistane-ketal *219* using endo-6-hydroxy-9-oxabicyclo[3.3.1]-nonan-2-one (*214*) or its acetate *221*, resp., was nevertheless successful by applying the following route: according to a procedure by Inhoffen et al.[66] the keto-acetate *221* was treated with trimethyl orthoformate. From the reaction mixture which contained the dimethoxy-ketals *222* and *223* as well as the enolethers *224* and *225*, the acetates *222* (63%) and *224* (5%) were isolated by chromatography. On base-hydrolysis they gave the alcohols *223* and *225*. Pyrolysis of the ketal *222* and simultaneous distillation yielded 67% of the enolether *224*. Finally by treating the dime-

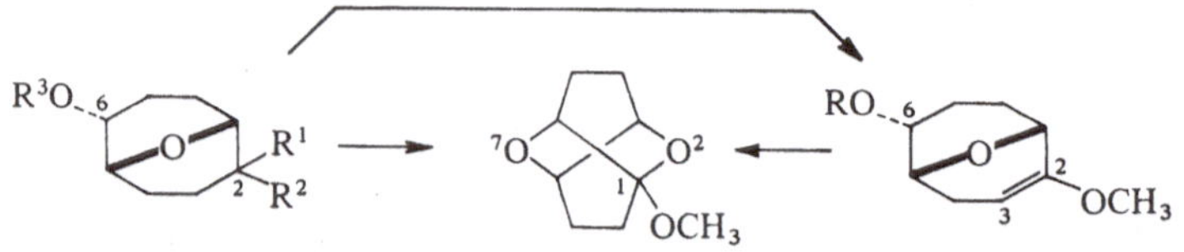

219

	R^1	R^2	R^3			R
214	O		H		224	Ac
221	O		Ac		225	H
222	OCH$_3$	OCH$_3$	Ac		226	CHO
223	OCH$_3$	OCH$_3$	H			

thoxy-ketal *223* with p-tolylsulfonic acid in benzene for 1 hr at 80° the twistane-ketal *219* was obtained in a 17% yield beside 57% of *214* (formed by hydrolysis) and other not identified products. Analogous treatment (under addition of trimethyl orthoformate to avoid hydrolysis by moisture) of the enolether *225* gave 66% of crude *219*, which contained approx. 10% of the formate *226*. The methoxy-twistane

219 was also already formed during the reaction of the hydroxy-ketone *214* with trimethyl orthoformate and acid. However, it is difficult to isolate *219* from this reaction mixture.

2.3.3. 2-Oxa-7-aza-twistanes (correct name) or 2-Aza-7-oxa-twistanes, Resp.[t)]

2.3.3.1. Pathway N (G 15 → G 32)[58)]

As already mentioned in 2.2.7. only traces of the twistane *203* were obtained by olefin amination with mercuric acetate in dimethyl sulfoxide of the bicyclic amine *201*. The main product was the corresponding 2-aza-7-oxa-isotwistane *202*.

2.3.3.2. Pathway O (G 30 → G 33)[58)]

Both bicyclic amines *227* and *228* with an exo-bromine atom at C(6) as the leaving group could be cyclized almost quantitatively in 1,2-dichloroethane in the presence

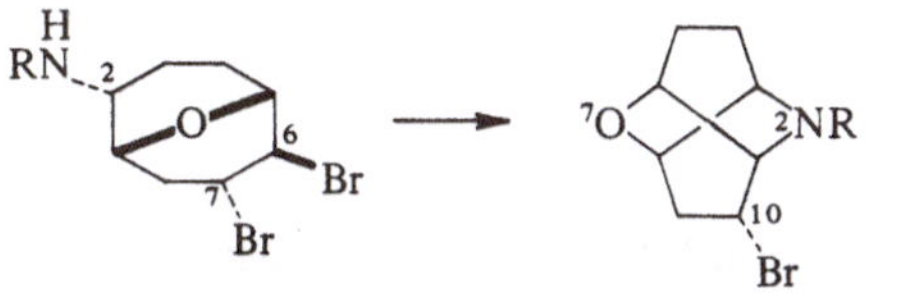

	R			R
227	H		*229*	H
228	CH_3		*230*	CH_3

of solid Na_2CO_3 to the $10^{N(2)}$-bromo-twistanes *229* and *230*. The former was converted to the latter by treatment with methyliodide (88%).

2.3.4. 2-Thia-7-aza-twistanes (correct name) or 2-Aza-7-thia-twistanes, Resp.[t)]: Pathway O (G 30 → G 33)[58)]

Studies on intramolecular cyclizations using pathway O were mainly carried out with the bicyclic endo-2-amine *231* having an exo-bromine atom at C(6) as leaving group.

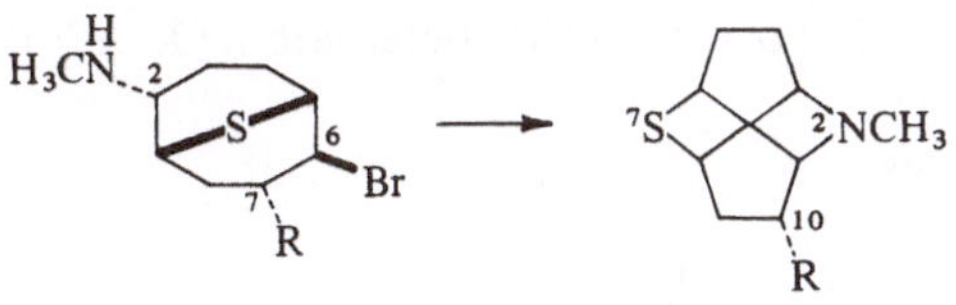

	R			R
231	Br		*234*	Br
232	Cl		*235*	Cl
233	OH		*236*	OH

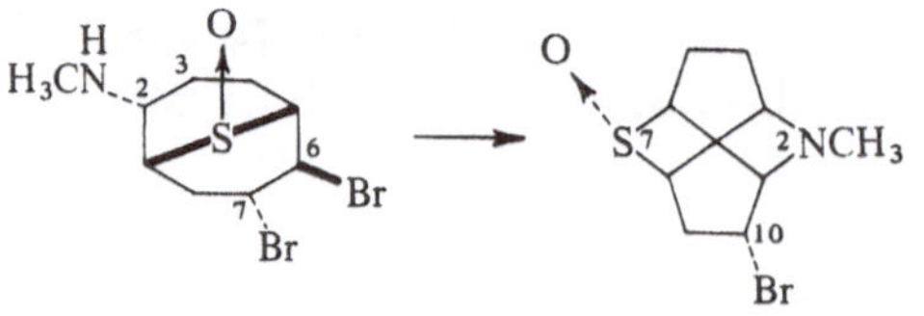

237   238

Best results have been obtained in 1,2-dichloroethane at 80° in the presence of solid Na_2CO_2, where the $10^{N(2)}$-bromo-2-aza-7-thia-twistane *234* was formed in 97% yield. Analogous cyclizations were carried out with *232* ($\rightarrow$ *235*), *233* ($\rightarrow$ *236*) and *237* ($\rightarrow$ *238*).

3. Syntheses: Substitutions and Rearrangements Involving Neighboring Group Participation of Dihetero-tricyclodecanes

3.1. Introduction

The following characterization of neighboring group participation was given by Capon[67]: "Some substituents may influence a reaction by stabilizing a transition state or intermediate by becoming bonded or partially bonded to the reaction centre. This behaviour is called neighbouring group participation, or sometimes, if an increased reaction rate results, intramolecular catalysis, and, as with intermolecular catalysis, nucleophilic, electrophilic, and basic catalysis or participation are possible. If the transition state of a rate-determining step is established in this way, an increased reaction rate results and the neighbouring group is then said to provide anchimeric assistance"[u].

In the series of dihetero-tricyclodecanes, substitutions and rearrangements involving neighboring group participation of heteroatoms (onium ions)[v], especially of oxygen (oxonium ions)[w], sulfur (episulfonium ions, thiiranium ions)[x, y] and nitrogen (ammonium ions, aziridinium ions)[z] as well as in one case also of selenium (episelenium ions), were studied on adamantanes (*G 1*), isotwistanes (*G 2*), twistanes (*G 3*) and homotwistbrendanes (*G 4*).

Compounds with one or two, resp., leaving groups at one or two, resp., β-carbon atoms in stereoelectronically favored positions (anti-periplanar) were chosen as candidates for neighboring group participation of one or the other heteroatom (X or Y):

one leaving group   at C(4)   in adamantanes
                     at C(10) in isotwistanes
                     at C(10) in twistanes[t]
                     at C(6)   in homotwistbrendanes[t]

two leaving groups   at C(4)   and C(8)   in adamantanes
                     at C(4)   and C(10) in isotwistanes
                     at (C4)   and C(10) in twistanes[t]
                     at C(4)   and C(6)   in homotwistbrendanes[t]

According to the possible configurations of the leaving groups, syn or anti to the hetereoatoms, the starting materials may be classified by the following three groups:
– the sole or both leaving groups anti to $X(2)$
– the sole or both leaving groups anti to Y
– one leaving group anti to $X(2)$, the other one anti to Y

3.2. The Sole (R^1) or Both (R^1 and R^2) Leaving Groups Anti to X(2)

3.2.1. Introduction

Depending on the number of leaving groups or substituents, resp., the listing in Table 1 follows for the starting materials.

Starting from the above compounds the following substitutions without and with skeletal rearrangements are possible. Intramolecular nucleophilic attack of the heteroatom $X(2)$ at the R^1-substituted carbon atom of adamantanes $G\ 36$ as well as of isotwistanes $G\ 37$ leads to the same onium ion $G\ 38$. A priori, an external nucleophile $R^{3\ominus}$ has the three possibilities A, B and C$^{aa)}$ for an attack at a carbon atom. Attack A yields $R^{3[Y(6)]}$-adamantanes $G\ 39$ (substitution under retention), attack B leads to $R^{3[Y(7)]}$-isotwistanes $G\ 40$ (substitution under skeletal rearrangement). Analogously adamantanes $G\ 42$ and isotwistanes $G\ 43$ may be obtained from adamantanes $G\ 39$ or isotwistanes $G\ 47$ through onium ions $G\ 41$ [formed by intramolecular attack of the heteroatom $X(2)$ at the R^2-substituted carbon atom] on the basis of the two possible attacks D and E$^{aa)}$.

Involving neighboring group participation of the heteroatom $X(2)$ by intramolecular nucleophilic substitution at the R^1-substituted carbon atom, an identical onium ion $G\ 46$ is formed starting from isotwistanes $G\ 44$ as well as from twistanes $G\ 45$.

Table 1

Starting materials		Leaving groups	
		One	Two
Adamantanes	$G\ 36$	R^1 (R^2 = H)	R^1 (R^2)
	$G\ 39$	R^2 (R^3 = substituent)	–
Isotwistanes	$G\ 37$	R^1 (R^2 = H)	R^1 (R^2)
	$G\ 40$	R^2 (R^3 = substituent)	–
	$G\ 44$	R^1 (R^2 = H)	R^1 (R^2)
	$G\ 47$	R^2 (R^3 = substituent)	–
Twistanes	$G\ 45$	R^1 (R^2 = H)	R^1 (R^2)
	$G\ 48$	R^2 (R^3 = substituent)	–

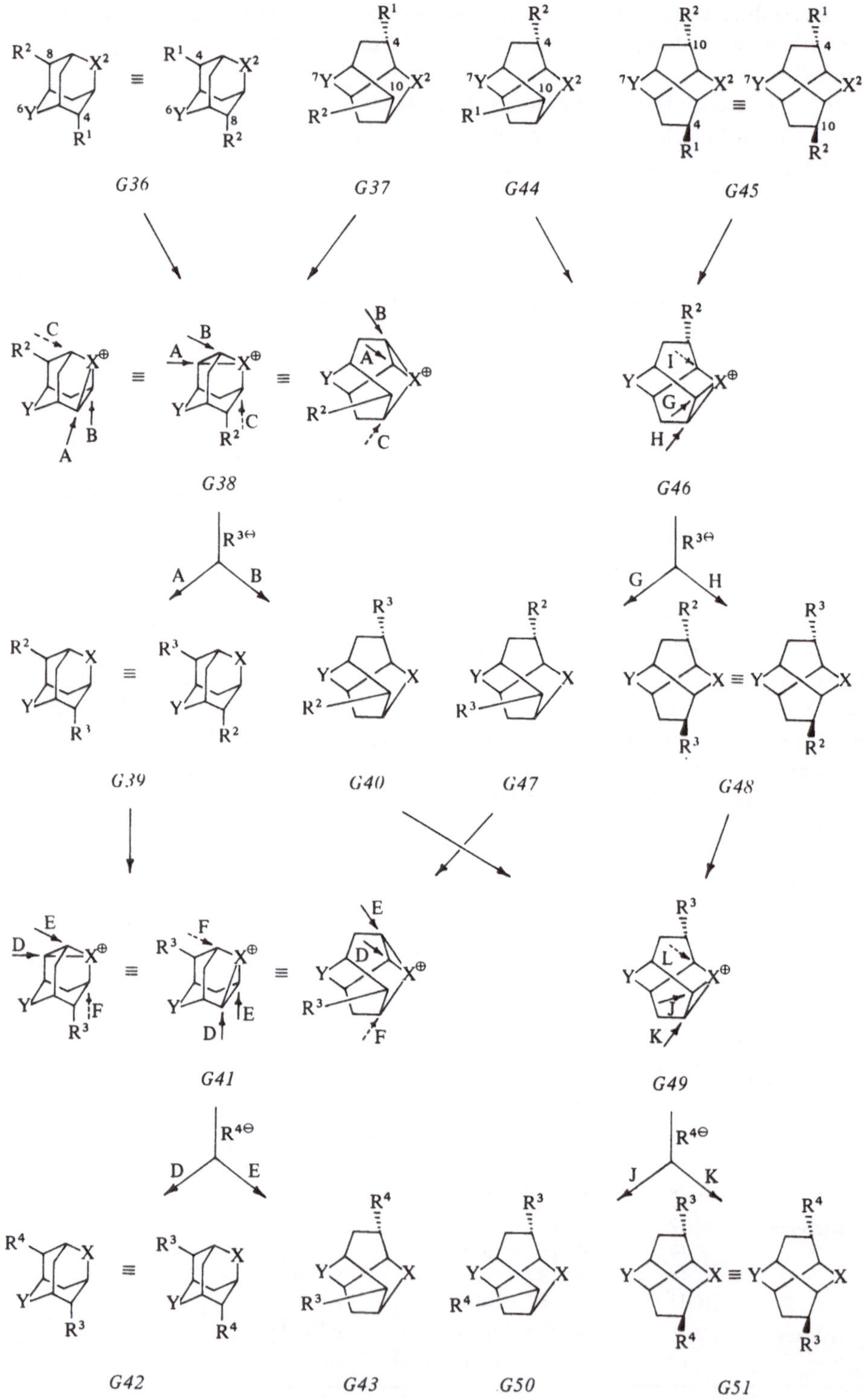

G36   *G37*   *G44*   *G45*

G38   *G46*

G39   *G40*   *G47*   *G48*

G41   *G49*

G42   *G43*   *G50*   *G51*

Of the a priori three possible attacks G, H and I[aa)] of an external nucleophile $R^{3\ominus}$ at *G 46*, G gives isotwistanes *G 47* and H twistanes *G 48*. Analogously the onium ion *G 49* [formed by attack of X(2) at the R^2-substituted carbon atom in isotwistanes *G 40* and twistanes *G 48* opens the access to isotwistanes *G 50* (attack J) and twistanes *G 51* (attack K)].

In the isotwistanes *G 40* and *G 47* on the one hand or in *G 43* and *G 50* on the other hand in each case only R^2 and R^3 or R^3 and R^4, resp., are interchanged. Applying the same external nucleophile ($R^{3\ominus} = R^{4\ominus}$) *G 43* and *G 50* become identical.

Several of the above mentioned possibilities of substitutions without and with skeletal rearrangements involving neighboring group participation were studied and are described as follows.

3.2.2. Starting Material: 2,6-Dihetero-adamantanes

Depending on the nature of the heteroatoms X and Y, 2,6-dihetero-adamantanes of the type *G 36* as starting materials where both leaving groups R^1-C(4) and R^2-C(8) are anti to the same heteroatom suffer substitutions either exclusively without skeletal rearrangement or together with rearrangement. However, rearranged products alone have not been obtained in any of the studied cases.

3.2.2.1. Only Substitution

Exclusively substitution without skeletal rearrangement under retention of configuration [see *G 36* → *G 38* (onium ion) → *G 39* → *G 41* (onium ion) → *G 42*] was the result of the following reactions:

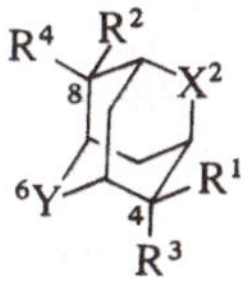

	Y(6)	X(2)	R^1	R^2	R^3	R^4
24	S	O	Cl	Cl	H	H
25	S	O	OH	OH	H	H
27	S	O	OCH_3	OCH_3	H	H
34	Se	O	Cl	Cl	H	H
35	Se	O	OAc	OAc	H	H
54	$NSO_2C_6H_5$	O	I	I	H	H
55	$NSO_2C_6H_5$	O	OAc	I	H	H
56	$NSO_2C_6H_5$	O	OAc	OAc	H	H
91	NCHO	S	H	H	Cl	Cl
92	NCHO	S	H	H	I	I
93	NCHO	S	H	H	OAc	OAc
94	$NSO_2C_6H_5$	S	H	H	Cl	Cl
95	$NSO_2C_6H_5$	S	H	H	OAc	OAc
116	NCH_3	NCH_3	H	H	OH	OH
118	NCH_3	NCH_3	H	H	Cl	Cl

— Treatment of $4^{O(2)},8^{O(2)}$-dichloro-2-oxa-6-thia-adamantane (*24*) with aqu. Na_2CO_3-solution yielded almost quantitatively the corresponding diol *25* (episulfonium) ion) (2.1.3.)[28], whereas reaction of *24* with silver oxide in methanol gave the dimethoxy-compound *27* (episulfomium ion) (2.1.3.)[21].

— Treatment of $4^{O(2)},8^{O(2)}$-dichloro-2-oxa-6-selena-adamantane (*34*) with silver acetate in acetic acid or acetonitrile almost quantitatively led to the corresponding diacetate *35* (episelenium ion) (2.1.4.)[27, 29].

— Using N(6)-formyl-$4^{N(6)},8^{N(6)}$-dichloro-2-thia-6-aza-adamantane (*91*) as starting material, reaction with sodium iodide in diethyl ketone yielded 82% of the diiodide *92* (episulfonium ion) and reaction with silver acetate in acetic acid gave 69% of the diacetate *93* (episulfonium ion) (2.1.9.)[40]. Under analogous conditions the dichloride *94* was transformed to the diacetate *95* (95%) (2.1.9.)[35].

— From treatment of N(2),N(6)-dimethyl-$4^{N(6)},8^{N(6)}$-dihydroxy-2,6-diaza-adamantane (*116*) with thionylchloride 35% of the corresponding dichloride *118* could be isolated (aziridinium ion) (2.1.10.3.)[44].

— A special case is the observed substitution with retention as the result of the reaction of N(6)-phenylsulfonyl-$4^{O(2)},8^{O(2)}$-diiodo-2-oxa-6-aza-adamantane (*54*) with silver acetate in acetic acid during 9 hrs at 60°, which yielded $4^{O(2)}$-acetoxy-$8^{O(2)}$-iodo-adamantane *55* (20% relative to reacted *54*) and $4^{O(2)},8^{O(2)}$-diacetoxy-adamantane *56* (66% relative to reacted *54*). Because of the reduced basicity of N(6) by the sulfone group this reaction will not be classified as one involving "normal" neighboring group participation (2.1.5.2.)[36].

3.2.2.2. Substitution and Rearrangement

In both cases studied using *G 36*-adamantanes as starting materials rearranged products were observed where oxonium ions occurred as intermediates, namely on treatment of $4^{O(6)},8^{O(6)}$-diiodo-2,6-dioxa-adamantane (*5*: 2.1.2.1.) and N(6)-phenylsulfonyl-$4^{N(6)},8^{N(6)}$-diiodo-2-oxa-6-aza-adamantane (*47*: 2.1.5.2.) with silver acetate in acetic acid, see *G 36* → *G 38* (oxonium ion) → *G 39* (adamantane) + *G 40* (isotwistane); *G 39* → *G 41* (oxonium ion) → *G 42* (adamantane) + *G 43* (isotwistane); *G 40* → *G 49* (oxonium ion) → *G 50* (isotwistane) + *G 51* (twistane).

	Y(6)	R^1	R^2		Y(7)	R		Y(7)
5	O	I	I	*242*	O	I	*243*	O
7	O	OAc	I	*23*	O	OAc	*245*	NR
9	O	OAc	OAc	*244*	NR	I		
47	NR	I	I	*78*	NR	OAc		
48	NR	OAc	I					
49	NR	OAc	OAc					$R = SO_2C_6H_5$

50

Table 2

Starting material G 36	Products (yields in %)							Refs.
	Total yield	G 36	G 39	G 40	G 42	G 43 = G 50	G 51	
5	94.5	5: 2.5	7: 19.5	242: –	9: 17.5	23: 34.5	243: 20.5	25)
47	67	47: 2	48: 28	244: –	49: < 1	78: 24	245: 13	35)

51

To obtain an equal turnover, the following reaction conditions had to be applied: 16 hrs at 95° with the 2,6-dioxa-adamantane *5* and 230 hrs at 70–75° with the 2-oxa-6-aza-adamantane *47*. The product distributions are listed in Table 2. It is remarkable that in no case the intermediate iodo-acetoxy-isotwistanes *242* and *244*, resp., could neither be isolated nor even be observed. The subsequent reactions to *23* + *243* and *78* + *245*, resp., are obviously faster than the formation of *242* from *5* and *244* from *47*. This is in agreement with the result that on the one hand $10^{O(7)}$-iodo-2,7-dioxa-isotwistane (*125*), i.e. a corresponding compound without an acetoxy group at C(4), on treatment with silver acetate in acetic acid shows complete interconversion already at room temperature (see 3.2.3.) and on the other hand that by analogous treatment of the $10^{N(7)}$-iodo-2-oxa-7-aza-isotwistane *169* after 3 hrs at 60° already no starting material was left (see 3.2.3.).

3.2.3. Starting Material: 2,7-Dihetero-isotwistanes

Isotwistanes of the general types *G 37* and *G 44* with two leaving groups R^1 and R^2 [at $C(4)^{Y(7)}$ and $C(10)^{Y(7)}$ or vice versa] as well as *G 47* with one leaving group R^2-$C(4)^{Y(7)}$ and one substituent R^3-$C(10)^{Y(7)}$ have not yet been used as starting materials. However, neighboring group participation of the heteroatom X(2) was studied starting from several isotwistanes of the type *G 44* with only one leaving group R^1-$C(10)^{Y(7)}$ (R^2 = H), see *G 44* → *G 46* (onium ion) → *G 47* (isotwistane) + *G 48* (twistane). The ratio of substitution without skeletal rearrangement (*G 47*) to substitution together with rearrangement (*G 48*) varies predominantly with the nature of the heteroatoms X(2) and Y(7) and only to a smaller extent also with the attacking nucleophile and the applied solvent applying comparable reaction temperatures. The results of the reactions of $10^{Y\ (7)}$-substituted isotwistanes (leaving group R^1 = J, Br, Cl, OH, OTs) are listed in Table 3. The highest yields (approx. 50%) in rearrangement products were obtained from isotwistanes with X(2) = O and Y(7) = O. The observed decrease of rearrangement can be summarized by the following sequences:

for X(2) = O: $\quad$ Y(7) $= O > NCO_2 C_2 H_5 > S$
for X(2) = S: $\quad$ Y(7) $= O > NCH_3 > S$
for X(2) = NR: $\quad$ Y(7) $= O > S$
for Y(7) = O; $\quad$ X(2) $= O > NCH_3 > S$
for Y(7) = S: $\quad$ X(2) $= O > NCH_3 > S$
for Y(7) = NR: $\quad$ X(2) $= O > S$

Isotwistanes of the type *G 40* [leaving group R^2-$C(10)^{Y(7)}$, R^3-$C(4)^{Y(7)}$ = substituent], although they are intermediates in the reactions of adamantanes *G 36*, are immediately further transformed under the applied reaction conditions: → *G 49* (onium ion) → *G 50* (isotwistane) + *G 51* (twistane) (see 3.2.2.2.). In isolated form such compounds *G 40* have not yet been used as starting materials.

3.2.4. Starting Material: 2,7-Dihetero-twistanes

Analogously to the isotwistanes (see 3.2.3.) twistanes with only one leaving group R^1-$C(10)^{Y(7)}$ (R^2 = H, type *G 45*)^t) were subjects for studies of neighboring group participation. The results are listed in Table 4.

Table 3

Starting material G 44	Y(7)	X(2)	R¹	Reagents	R³	Isotwistane G 47	Ratio approx.	Twistane[t] G 48	Refs.
125	O	O	I	AgOAc/AcOH	OAc	246	1 : 1	247	45, 50)
147	S	O	I			248	3.3 : 1	249	53)
164	$O-S(7)^{C(4)bb}$	O	I			250	2.5 : 1	251	54)
163	SO_2	O	I			252	6 : 1	253	54)
169	$NCO_2C_2H_5$	O	I			254	2 : 1	255	55)
183	O	S	Br			256	5 : 1	257[t]	27)
258	O	S	Cl			256	9 : 1	257[t]	27)
259	NCH_3	S	Cl	$AgOAc/CH_2Cl_2$		260	30 : 1	261	58)
198	S	S	Br	AgOAc/AcOH		262	> 50 : 1	263	57)
199	S	S	Cl			262	> 50 : 1	263	57)
264	O	NCH_3	Cl	$AgOAc/CH_2Cl_2$		202	2.5 : 1	203[t]	58)
265	S	NCH_3	Cl			206	7.7 : 1	266[t]	58)
125	O	O	I	$AgOTs/CH_3CN$	OTs	267	3 : 2	268	50)
147	S	O	I			269	> 100 : 1	270	53)
169	$NCO_2C_2H_5$	O	I			271	63 : 1	272	55)
183	O	S	Br	Na_2CO_3/H_2O	OH	273	5.7 : 1	274[t]	27)
198	S	S	Br			275	> 50 : 1	276	57)
199	S	S	Cl			275	> 50 : 1	276	57)

Table 3 (continued)

Starting material G 44	Y(7)	X(2)	R¹	Reagents	R³	Isotwistane G 47	Ratio approx.	Twistane[t] G 48	Refs.
275	S	S	OH	SOCl₂	Cl	199	>50:1	277	27)
278	NCH₃	S	OH			259	>100:1	279	58)
280	O	NCH₃	OH			264	>100:1	281[t]	58)
282	S	NH	OH			283	>100:1	284[t]	58)
285	S	NCH₃	OH			265	>100:1	286[t]	58)
267	O	O	OTs	NaN₃/HMPT	N₃	287	1:1	288	50)
267	O	O	OTs	LiAlH₄	H	128	1.6:1	212	50)
269	S	O	OTs			148	2.8:1	289	53)
267	O	O	OTs	LiAlD₄	D	290	–	291	24)
269	S	O	OTs			292	2.8:1	293	53)
198	S	S	Br	LiAlH₄	H	200	>100:1	294	57)
199	NCH₃	S	Cl			200	>100:1	294	57)
259	O	S	Cl			295	40:1	296	58)
264	O	NCH₃	Cl			297	2.8:1	298[t]	58)
265	S	Cl	NCH₃			299	8:1	296[t]	58)

Table 4

Starting material $G\,45$	$Y(7)$	$X(2)$	R^1	Reagents	R^3	Isotwistane $G\,47$	Ratio approx.	Twistane $G\,48$	Refs.
268	O	O	OTs	LiAlH$_4$	H	128	1.6 : 1	212	50)
268	O	O	OTs	LiAlD$_4$	D	290	1.6 : 1	291	24)
270	S	O	OTs	LiAlH$_4$	H	148	2.8 : 1	289	53)
270	S	O	OTS	LiAlD$_4$	D	292	2.8 : 1	293	53)
300^t	O	NCH$_3$	OH	SOCl$_2$	Cl	264	$\geqslant$ 100 : 1	281^t	58)
301^t	S	NCH$_3$	OH	SOCl$_2$	Cl	265	$\geqslant$ 100 : 1	286^t	58)
257^t	O	S	OAc	ΔT	OAc	256	$\geqslant$ 100 : 1	257^t	27)
266^t	S	NCH$_3$	OAc	ΔT	OAc	206	$\geqslant$ 100 : 1	266^t	58)
302				H$_2$SO$_4$/H$_2$O	OH	278	$\geqslant$ 100 : 1	303	58)

Of special importance are the results of the LiAlH$_4$- and LiAlD$_4$-reductions of the tosylates *268* and *270*. The product-ratios of isotwistane to twistane *128* to *212* = 1.6 : 1 and *148* to *289* = 2.8 : 1, resp., are exactly the same as the ones obtained starting from the isotwistanes *267* and *269* (see 3.2.3.). This proves the existence of a common intermediate, an oxonium ion of the general type *G 46*. A further hint for its being an intermediate gave the stereospecific deuterium incorporation D-C(10)$^{Y(7)}$ in the LiAlD$_4$-reduction products of *268* and *270*. Mixtures of *290* and *291* (1,6 : 1) and *292* and *293* (2.8 : 1), resp., were obtained, each with 99% d$_1$, again the same results as starting from the corresponding isotwistanes *267* and *269* (see 3.2.3.).

The pyrolysis experiments with the twistanes *257*[t] and *266*[t] reflect the relative stabilities of twistanes and isotwistanes, the latter being strongly favored.

An interesting reaction is the addition of water to the twistane *302*. The obtained product, exclusively 10$^{N(7)}$-hydroxy-2-thia-7-aza-isotwistane (*278*), proved the expectation to be correct that sulfur will predominate over nitrogen in neighboring group participation. Furthermore, experiments with D$_2$SO$_4$ in D$_2$O allowed a dif-

	R^1	R^2		R^1	R^2
G 52	D	H	*304*	D	H
G 53	H	D	*305*	H	D

ferentiation between the two possible pathways involving either the episulfonium ion *G 52* (→ *304*) or *G 53* (→ *305*). The C(9)$^{N(7)}$-deuterated alcohol *305* was the sole product after exchange of DO-C(10) to HO-C(10).

3.3. The Sole (R^1) or Both (R^1 and R^2) Leaving Groups Anti to Y

3.3.1. Starting Material: 2,6-Dihetero-adamantanes

For symmetry reasons adamantanes with the sole (R^1) or both (R^1 and R^2) leaving groups anti to the heteroatom Y(6) were classified and therefore already discussed in 3.2.2. as compounds of the general type *G 36* (exchange of the two heteroatoms).

3.3.2. Starting Material: 2,7-Dihetero-isotwistanes and -twistanes as Well as 2,8-Dihetero-homotwistbrendanes

With respect to the orientation of the leaving group no isotwistanes, twistanes and homotwistbrendanes (see 3.1.) are known so far as having both leaving groups anti to the heteroatom Y, which could exhibit neighboring group participation.

3.4. One Leaving Group (R^1) Anti to X(2), the Other (R^2) Anti to Y

3.4.1. Introduction

Compounds with one leaving group (R^1) anti to X(2) and a second (R^2) anti to Y[cc)] like adamantanes G 54 [R^1-C(4)$^{Y(6)}$, R^2 -C(8)$^{X(2)}$], isotwistanes G 55 [R^1 -C(4)$^{Y(7)}$, R^2 -C(10)$^{X(2)}$] and homotwistbrendanes G 56 [R^1 -C(4)$^{Y(8)}$, R^2 -C(6)$^{X(2)}$] have two possibilities for neighboring group participation either involving X(2) or Y (see the general scheme[dd)]).

Using an identical nucleophile (R$^{3\ominus}$ = R$^{4\ominus}$) the disubstituted products G 73 and G 77 (adamantane), G 74 and G 79 [isotwistane: X(2), Y(7)], G 75 and G 78 [isotwistane: Y(2), X(7)] as well as G 76 and G 80 (homotwistbrendane) are identical, too. So far, only adamantanes (G 54) and isotwistanes (G 55) as starting materials have been the subject of studies.

3.4.2. Starting Material: 2,6-Dihetero-adamantanes

Table 5 summarizes the results of the treatment of 4$^{O(6)}$,8$^{O(2)}$-diiodo-2,6-dioxa-adamantane (6) with silver acetate in acetic acid. Because of the identity of X(2) and Y(6), in 8 compounds of the general types G 63 and G 61 (corresponding onium ions G 58 and G 57, resp.), in 304 such of the types G 64 and G 62 (corresponding onium ions G 58 and G 57, resp.), in 10 such of G 77 and G 73 (corresponding onium ions G 71 and G 69, resp.), in 306 such of G 80 and G 76(corresponding onium ions G 72 and G 70, resp.) as well as in 305 even such of G 78, G 75, G 79 and G 74 (corresponding onium ions G 71, G 70, G 72 and G 69, resp.) are identical.

The product distributions, especially the almost equal amount of 304 obtained under both reaction conditions applied, indicate that the diacetoxy-isotwistane 305 was predominantly formed through the iodo-acetoxy-adamantane 8 because of the obviously smaller reactivity of the 10$^{O(2)}$-iodo-isotwistane 304. It has to be noted that a homotwistbrendane 306 could not be detected.

A further example was the reaction of N(6)-phenylsulfonyl-4$^{N(6)}$,8$^{O(2)}$-diiodo-2-oxa-6-aza-adamantane (50) with silver acetate (Table 6). The 4$^{N(6)}$-acetoxy-8$^{O(2)}$-

Table 5[25)]

Starting material	Reaction conditions	Products [%]				
		6	8	304	10	305
6	1 mol-equ. AgOAc 48 hrs 70°	28	14.5	22	18	
6	4 mol-equ. AgOAc 15 hrs 95°	1.5	–	18	71 (21 : 79)	

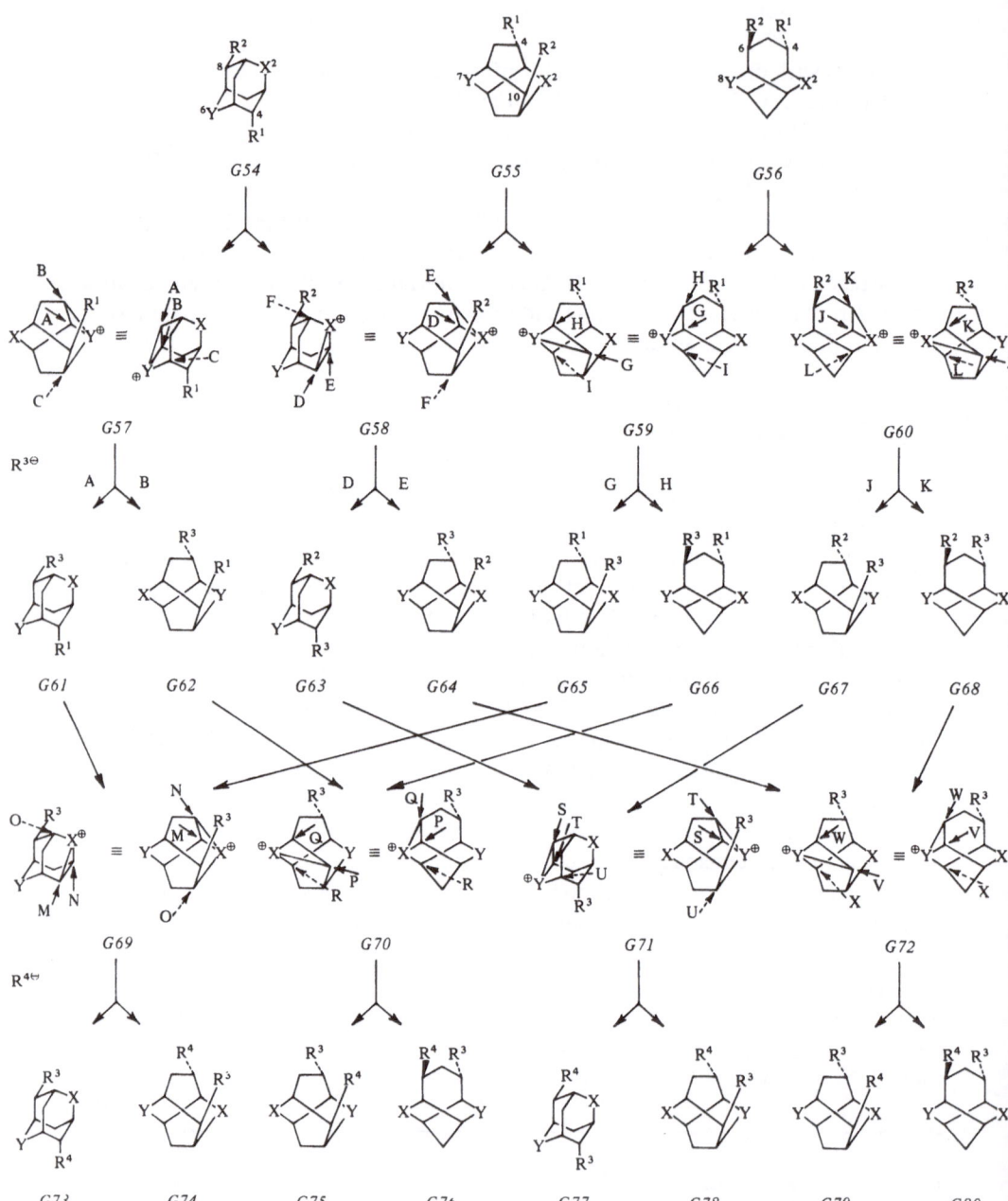

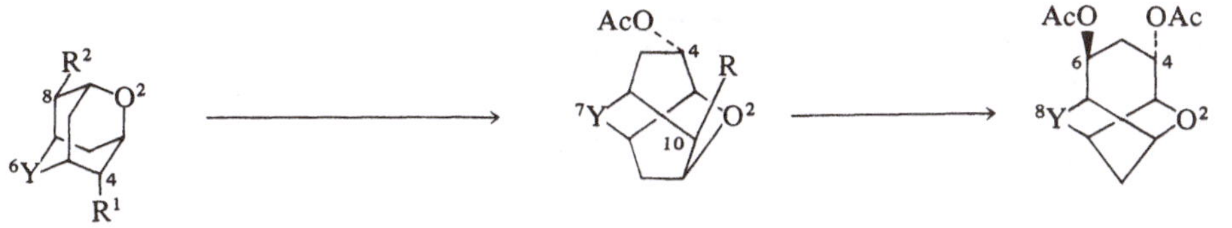

	Y(6)	R¹	R²		Y(7)	R		Y(7)
6	O	I	I	304	O	I	306	O
8	O	OAc	I	305	O	OAc	309	NR³
10	O	OAc	OAc	307	NR³	I		
50	NR³	I	I	308	NR³	OAc		
51	NR³	I	OAc					
52	NR³	OAc	I					
53	NR³	OAc	OAc					

$$R^3 = SO_2C_6H_5$$

iodo-adamantane *52* could neither be isolated nor detected. The results demonstrate that in the adamantane *50*, $I-C(8)^{O(2)}$[anti to N(6)] is much more reactive than $I-C(4)^{N(6)}$ [anti to O(2)] (ratio approx. 8 : 1). On the other hand comparing the product distributions after 4 and 89 hrs indicate that the iodine $I-C(4)^{N(6)}$ [anti to O(2)] in the iodo-acetoxy-adamantane *51* is more reactive than $I-C(10)^{O(2)}$ [anti to N(7)] in the iodo-acetoxy-isotwistane *307*. This follows that the diacetates *53*

Table 6[35)]

Starting material	Reaction conditions	Products [%]					
		50	52	307	51	53	308
50	excess AgOAc 4 hrs 70°	–	–	10	77	–	–
50	excess AgOAc 89 hrs 70°	–	–	7	50	20	10

(adamantane) and *308* (isotwistane) are predominantly obtained from *51*. Therefore the reactions involving onium ions correspond to G *54* (adamantane) → G *58* (oxonium ion) → G *64* (isotwistane) → G *72* (onium ion) → G *79* (isotwistane) and to G *61* (adamantane) → G *69* (oxonium ion) → G *73* (adamantane) + G *74* (isotwistane). However, because of $R^3 = R^4$, G *74* and G *79* can not be differentiated, they are identical. As in the above case of the dioxa-compounds no homotwistbrendane *309* could be detected.

Table 7

Starting material G 55	Y(7)	X(2)	R²	Reagents	R³	Isotwistane G 65	Ratio approx.	Homotwist-brendane[t] G 66	Refs.
127	O	O	I	AgOAc/AcOH	OAc	310	11 : 1	311	50)
146	S	O	I	NaOAc/AcOH	OAc	312	>100 : 1	313	53)
145	S	O	Br	NaOAc/AcOH	OAc	312	>100 : 1	313	53)
146	S	O	I	AgOTs/CH$_3$CN	OTs	314	>100 : 1	315	53)
314	S	O	OTs	LiAlH$_4$/Dioxan	H	148	>100 : 1	316	53)
317	S	O	OH	p-TsCl/HCl	Cl	318	>100 : 1	319	53)
170	NCO$_2$C$_2$H$_5$	O	I	AgOAc/AcOH	OAc	320	>100 : 1	321	55)
322	O	S	I	AgOAc/AcOH	OAc	323	>100 : 1	324[t]	27)
325	S	S	Cl	AgOAc/AcOH	OAc	326	>100 : 1	327	27)
328	S	S	OH	SOCl$_2$/CHCl$_3$	Cl	325	>100 : 1	329	27)

3.4.3. Starting Material: 2,7-Dihetero-isotwistanes

Neighboring group participation of Y(7) in isotwistanes was studied on several compounds of the type $G\ 55$ with only one leaving group R^2-$C(10)^{X(2)}$ (R^1 = H). The results are summarized in Table 7. Two types of products can be formed: isotwistanes $G\ 65$ and homotwistbrendanes $G\ 66$. Only in the dioxa-series compounds of the latter type $G\ 66$ could be found. Treatment of $10^{O(2)}$-iodo-2,7-dioxa-isotwistane (127) with silver acetate yielded 8% of $6^{O(2)}$-acetoxy-2,8-dioxa-homotwistbrendane (311)[t], the main product (92%) being the isotwistane-acetate 310. In all other examples with only one or no oxygen bridge no homotwistbrendanes $G\ 66$ could be observed.

322

323

256

G81   G82

Special attention should be paid to the result obtained by treating $10^{S(2)}$-iodo-2-thia-7-oxa-isotwistane (322) with silver acetate. The corresponding acetoxy-isotwistane 323 was the main product, however, the $10^{O(7)}$-acetoxy-2-thia-7-oxa-isotwistane (256) was also formed, the ratio of the products being approx. 3.5 : 1. This clearly proves that the primarily formed oxonium ion $G\ 81$ was partially converted to the isomeric episulfonium ion $G\ 82$ in competition with the attack of the external nucleophile[27].

4. Syntheses: Further Transformations and Derivatives of Dihetero-tricyclodecanes

4.1. Introduction

Starting from dihetero-tricyclodecanes described in the Sections 2. and 3. of the types of adamantane ($G\ 1$), isotwistane ($G\ 2$), twistane ($G\ 3$) and homotwistbrendane ($G\ 4$), which in their turn were prepared either by cyclization of suitable 9-heterobicyclo-[3.3.1]-and 9-heterobicyclo[4.2.1]nonanes or by substitution without or together with rearrangement of already available dihetero-tricyclodecanes, several further derivatives have been synthesized.

4.2. 2,6-Dihetero-adamantanes

All the known derivatives of 2,6-dihetero-adamantanes have already been described in the Sections 2.1., 3.2.2., 3.3.1. and 3.4.2.

61

4.3. 2,7-Dihetero-isotwistanes

4.3.1. 2,7-Dioxa-isotwistanes[45, 50]

Base-hydrolysis of the $10^{O(7)}$-acetoxy-isotwistane 246 (3.2.3.) yielded the corresponding alcohol 137, which was also characterized as its tosylate 267 (3.2.3.). The azide
287 (3.2.3.) was catalytically reduced (H_2/raney-nickel) to the $10^{O(7)}$-amino-iso-
twistane 330. Oxidation of the isotwistanol 137 with Jones-reagent gave the ketone
331, which was transformed via desulfuration with raney-nickel of the corresponding
thioketal 332 to unsubstituted 2,7-dioxa-isotwistane (128, see also 2.2.2.1.).

	R		R		R	
246	OAc	331	O	310	OAc
137	OH			333	OH
267	OTs	332	S—S (ketal)	334	OTs
287	N$_3$				
330	NH$_2$	128	H$_2$		

Base-hydrolysis of the $10^{O(2)}$-acetoxy-isotwistane 310 (3.4.3.) gave the alcohol
333, which was also obtained as sole product in a 99% yield by LiAlH$_4$-reduction
of the ketone 331. LiAlH$_4$-treatment of the tosylate 334, prepared from the alcohol
310, yielded only alcohol 310 again and no unsubstituted 2,7-dioxa-isotwistane (128).
Obviously the cleavage of the bond between O–C(10)$^{O(2)}$ and S–OC(10)$^{O(2)}$ is
preferred over the formation of an oxonium ion G 59. This result, compared with
the exclusive formation of an oxonium ion G 46 (no reaction to the alcohol 137)
in the analogous treatment of the $10^{O(7)}$-tosylate 267 (3.2.3.), is a qualitative mea-
sure for the differences of the tendencies for the formation of the oxonium ions
G 46 and G 59, starting from C(10)$^{O(7)}$- and C(10)$^{O(2)}$-substituted isotwistanes[ee].
In agreement with that is the observed difference in reactivity of the iodides 125
and 127 when treated with silver acetate: the $10^{O(7)}$-iodide 125 reacts already at
room temperature ($\rightarrow$ 246 + 247, 3.2.3.), whereas the $10^{O(2)}$-iodide 127 only at reflux
temperature ($\rightarrow$ 310 + 311, 3.4.3.). These differences may be caused by the different
ring strains in the two oxonium ions G 46 and G 59. G 46 contains beside the new
three-membered ring (C1) –O(2) –C(10)[ff] an additional six-membered ring C(10)–
O(2) –C(3) –C(4) –C(5) –C(6)[ff], whereas G 59 contains beside the new three-membered ring C(6) –O(7) –C(10)[ff] only a five-membered ring C(10) –O(7) –C(8) –C(9) –
C(1)[ff].

4.3.2. 2-Oxa-7-thia-isotwistanes[53, 54]

Base-hydrolysis of $10^{S(7)}$-acetoxy-isotwistane 248 (3.2.3.) yielded the corresponding
alcohol 335, which was transformed into the tosylate 269 (see also 3.2.3.). The al-

cohol *335* was oxidized with CrO_3/pyridine to the ketone *336* (72%), which by Wolff-Kishner-reduction yielded unsubstituted 2-oxa-7-thia-isotwistane (*148*, see among others 2.2.3.1.).

Base-hydrolysis of the $10^{O(2)}$-acetoxy-isotwistane *312* (3.4.3.) gave the corresponding alcohol *317*, which could be oxidized by CrO_3/pyridine to the ketone *336* (85%). $LiAlH_4$-reduction of the latter led to a mixture of the two C(10)-epimeric alcohols *335* and *317* (ratio approx. 4 : 1).

	R			R				R
338	OAc		*248*	OAc	$\longrightarrow$	*250*		OAc
337	OH	$\longleftarrow$	*335*	OH				
			269	OTs				
155	HgI	$\longleftarrow$	*144*	HgI				
			147	I	$\longrightarrow$	*164*		I

	R			R			R
336	O		*312*	OAc		*339*	OAc
148	H_2		*317*	OH		*340*	H
			148	H			

Some 10-substituted 2-oxa-7-thia-isotwistanes were oxidized with one equivalent of hydrogen peroxide in acetic acid for chemical correlations. In each case stereo-specifically[54] only one of the two possible S(7)-epimeric sulfoxides was formed. The $10^{S(7)}$-acetate *248* and $10^{S(7)}$-iodide *147* gave the corresponding sulfoxides *250* and *164* with $S(7)^{C(4)}$-configuration[bb], whereas the $10^{S(7)}$-alcohol *335* and the $10^{S(7)}$-iodomercuri compound *144* led to the sulfoxides *337* (also characterized as its acetate *338*) and *155* with $S(7)^{C(1)}$-configuration[bb]. On the other hand, oxidation of the sterically unhindered $10^{O(2)}$-acetate *312* as well as of the unsubstituted 2-oxa-7-thia-isotwistane (*148*) yielded the sulfoxides *339* and *340* also with $S(7)^{C(1)}$-configuration[bb].

4.3.3. 2-Oxa-7-aza-isotwistanes

All the derivatives known are described in the Sections 2.2.4. and 3.

4.3.4. 2-Thia-7-oxa-isotwistanes[27]

$10^{O(7)}$-Isotwistanol 273 (3.2.3.) was oxidized with pyridine-SO_3/dimethyl sulfoxide to the ketone 341 (60%). $NaBH_4$-reduction in 1,2-dimethoxyethane yielded exclusively the $10^{S(2)}$-alcohol 342, which was acetylated ($\rightarrow$ 323, see also 3.4.3.) and tosyl-

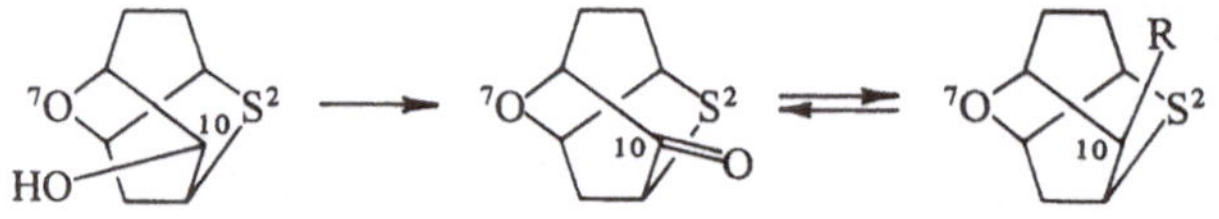

	R
342	OH
323	OAc
343	OTs
322	I

ated ($\rightarrow$ 343). The latter on $LiAlH_4$-treatment in refluxing dioxane yielded again the corresponding alcohol 342 as did the analogous compound in the 2,7-dioxa-series (334 $\rightarrow$ 333: 4.3.1.). Reaction of the tosylate 343 with magnesium iodide in ether at 80° in a sealed tube gave the $10^{S(2)}$-iodide 322, starting material for studies of neighboring group participation (see 3.4.3.).

4.3.5. 2,7-Dithia-isotwistanes[27, 29]

Collin's-oxidation of the $10^{S(7)}$-isotwistanol 275 (3.2.3.) yielded ketone 344, which on $LiAlH_4$-reduction gave exclusively the $10^{S(2)}$-isotwistanol 328 (87%), the starting material for the reaction with thionylchloride ($\rightarrow$ 325: 3.4.3.)[27].

An interesting result was obtained in the oxidation of 275 with one equivalent of hydrogen peroxide. Although the main product (60%) was the $S(2)^{C(9)}$-sulfoxide[gg]

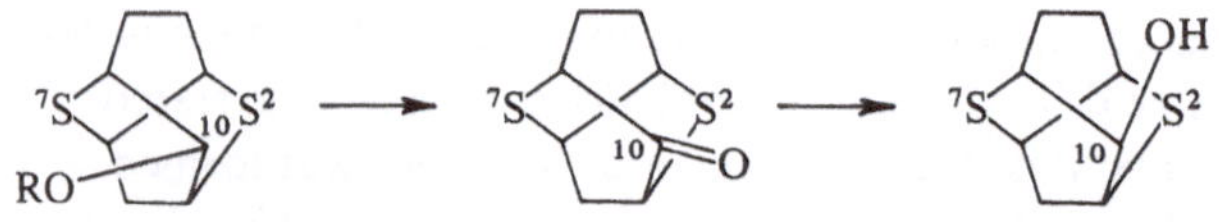

	R
275	H
262	Ac

	R
345	H
346	Ac

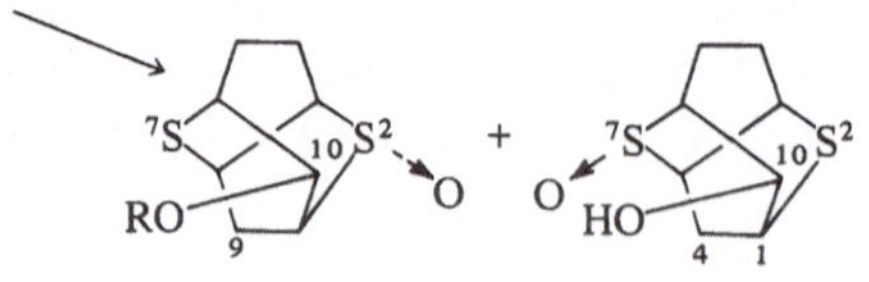

347

345 (attack from the most unhindered side) still 19% of the $S(7)^{C(1)}$-oxide[bb] *347* was isolated, whereas analogous oxidation of the corresponding acetate *262* yielded almost exclusively the $S(2)^{C(9)}$-sulfoxide[gg] *346*[29].

4.3.6. 2-Thia-7-aza-isotwistanes

All 2-thia-7-aza-isotwistanes were prepared from the $10^{N(7)}$-alcohol *278* (3.2.3.), which was obtained by addition of water to the twistane *302* (3.2.4.).

4.3.7. 2-Aza-7-oxa-isotwistanes

All the derivatives known are described in the Sections 2.2.7., 3.2.3. and 3.2.4.

4.3.8. 2-Aza-7-thia-isotwistanes[58]

The $10^{S(7)}$-acetates *206* and *207* (2.2.8.1.) were the starting materials for most of the derivatives prepared. Treatment of the N(2)-methyl-compound *206* with dimethyl

	R^1	R^2		R
206	CH$_3$	OAc	*350*	CH$_3$
285	CH$_3$	OH	*351*	H
207	H	OAc	*352*	CO$_2$C$_2$H$_5$
282	H	OH		
348	CO$_2$C$_2$H$_5$	OAc		
349	CO$_2$C$_2$H$_5$	OH		

sulfoxide at 130° yielded 46% of the oxidation product ketone *350* and 22% of the alcohol *285*. The latter, which was used as starting material for reactions involving neighboring group participation (see 3.2.3.), was also obtained quantitatively by base-hydrolysis of the acetate *206*.

Because of low stability and difficulties in separation from dimethyl sulfoxide after its preparation, the crude sec. amine *207* was immediately treated with ethyl chloroformate yielding N(2)-ethoxycarbonyl-isotwistane-acetate *348* [which could also be prepared from *206* (98%)] and some ketone *352*. The latter most probably was obtained from the ketone *351*, which in its turn was formed by oxidation of *207* with dimethyl sulfoxide. Base-hydrolysis of the acetate *348* gave the alcohol *349* (98%), which could also be obtained from the N(2)-methyl-isotwistanol *285* by reaction with ethyl chloroformate (87%). Treatment of the crude reaction mixture of the preparation of HN(2)-isotwistane-acetate *207* (2.2.8.1.) with a potassium carbonate-solution resulted in the formation of the alcohol *282*, which could also be prepared by cleavage of the N(2)-ethoxycarbonyl-isotwistanol *349* in 48% HBr/H_2O-solution (94%). *282* was also used as starting material for further reactions (see 3.2.3.).

4.4. 2,7-Dihetero-twistanes[t)]

4.4.1. 2,7-Dioxa-twistanes[25, 45, 50, 61)]

Base-hydrolysis of the $10^{O(7)}$-acetoxy-twistane *247* (3.2.3.) yielded the corresponding alcohol *353*, which was transformed to its tosylate *268* (3.2.3.) and mesylate *354*. Both compounds on treatment with potassium t-butoxide in dimethyl sulfoxide were

359

	R			R
247	OAc	*356*	O	
353	OH	*357*	S—S	
268	OTs			
354	OMs	*212*	H_2	
288	N_3	*358*	NOH	
355	NH_2			

331

360

easily converted to the twistene *360* (80%), which was the first example of a hetero-twistene (pure carbocyclic twistene had already been synthesized earlier[80)]). Catalytic hydrogenation (H_2/Pt or Pd, C) of the twistene *360* gave quantitatively 2,7-dioxa-twistane (*212*) (2.3.2.1.). Treatment of the tosylate *268* in dimethyl sulfoxide alone, however, yielded no twistene *360*, but the rearranged isotwistanone *331* (4.3.1.). The alcohol *353* was also oxidized with Jones-reagent to the ketone *356*, which by desul-furation with raney-nickel of the corresponding thioketal *357* yielded again unsub-stituted 2,7-dioxa-twistane (*212*).

10-Amino-twistanes as further derivatives were prepared in the following ways: catalytic reduction (H_2/raney-nickel) of the $10^{O(7)}$-azide *288* (3.2.3.) gave the primary $10^{O(7)}$-amino-twistane *355*. A separable mixture of the two C(10)-epimeric amino-twistanes *355* ($10^{O(7)}$-amine) and *359* ($10^{O(2)}$-amine) (ratio 43 : 57) was obtained by catalytic reduction (H_2/raney-nickel) of the oxime *358*, which was prepared from the ketone *356*.

2,7-Dioxa-twista-4,9-diene (*363*), which still represents the first heterocyclic twistadiene prepared thus far (meanwhile also pure carbocyclic twistadiene has been synthesized[81])) was easily accessible using the diacetate *243* (3.2.2.2.) as starting material. LiAlH$_4$-reduction yielded the diol *361* (90%), which was transformed to its dimesylate *362* (85%). Treatment of the latter for 3 days at room temperature with t-BuOK in dimethyl sulfoxide gave 70% of 2,7-dioxa-twista-4,9-diene (*363*), which was catalytically reduced (H$_2$/Pd, C) to the twistane *212* (85%).

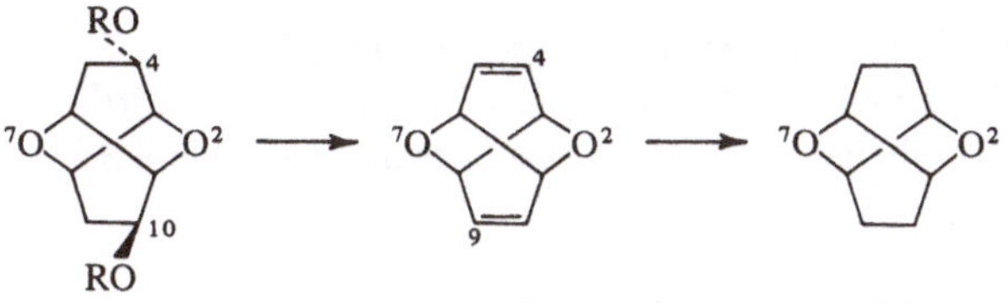

	R
243	Ac
361	H
362	Ms

4.4.2. 2-Oxa-7-thia-twistanes[53, 54)]

Base-hydrolysis of the 10$^{S(7)}$-acetoxy-twistane *249* (3.2.3.) yielded the alcohol *364*, which on the one hand was converted to the tosylate *270* (starting material for reactions discussed in 3.2.3.) and on the other hand oxidized (CrO$_3$/pyridine) to the

	R			R
249	Ac		*365*	O
364	H		*289*	H$_2$
270	Ts			

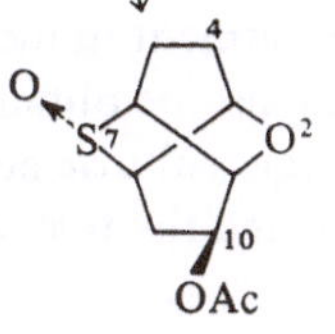

ketone *365*. Treatment of the latter under Wolff-Kishner conditions led to a mixture (approx. 65%) of endo-2-mercapto-9-oxabicyclo[3.3.1]non-6-ene (*178*) and -non-7-ene (*190*) in a ratio of approx. 1 : 5 and only traces (1–2%) of unsubstituted 2-oxa-7-thia-twistane (*289*: 3.2.3.). Oxidation of the acetate *249* with one equivalent of hydrogen peroxide gave the $S(7)^{C(4)}$-sulfoxide[bb] *251* (3.2.3.).

4.4.3. 2-Oxa-7-aza-twistanes (correct name) or 2-Aza-7-oxa-twistanes, Resp.[t] [55, 58]

Most of the derivatives have already been described in the Sections 2.3.3., 3.2.2.2., 3.2.3. and 3.2.4. It has to be added that both bromo-compounds *229*[t] and *230*[t] (2.3.3.2.) were easily converted to the corresponding twistenes *366* (83%) and *367*

	R			R			R^1	R^2
229	H		*366*	H		*255*	$CO_2C_2H_5$	Ac
230	CH_3		*367*	CH_3		*368*	CH_3	H
						369	CH_3	Ac

(86%), resp., on treatment with t-BuOK in tetrahydrofuran and that $LiAlH_4$-reduction of the N(7)-ethoxycarbonyl-acetate *255* (3.2.3.) led to the N(7)-methyl-alcohol *368*, which was also characterized as its acetate *369*.

4.4.4. 2-Thia-7-aza-twistanes (correct name) or 2-Aza-7-thia-twistanes, Resp.[t] [58]

One of the most suitable starting materials for the preparation of further derivatives of 2-aza-7-thia-twistanes[t] is the $10^{N(2)}$-bromide *234* (2.3.4.). Oxidation with one equivalent of hydrogen peroxide gave a mixture of the two sulfoxides *238* (19%) and *372* (38%). The former was identical with the one obtained by cyclization of the bicyclic compound *237* (2.3.4.). This allows conclusive assignment of the orientation of the oxygen atoms in the two sulfoxides *238* and *372*.

Treatment of the bromide *234* with t-BuOK in dimethyl sulfoxide yielded 65%, with t-BuOK in tetrahydrofuran even 81% of the twistene *302*, a starting material for the synthesis of 2-thia-7-aza-isotwistanes (3.2.4.).

Reaction of the bromide *234* with thionylchloride led to the corresponding chloride *235* (76%) (2.3.4.). Both halides *234* and *235* with silver acetate in acetic acid yielded the acetate *370* (58% and 75%, resp.), which on treatment in methanolic KOH-solution gave the alcohol *236* (2.3.4.). The latter can also be obtained directly (80%) by warming a solution of the bromide *234* in 20% aqu. sulfuric acid.

Treatment of the bromide *234* with silver tosylate yielded the tosylate *371* (71%), which can also be prepared from the alcohol *236*.

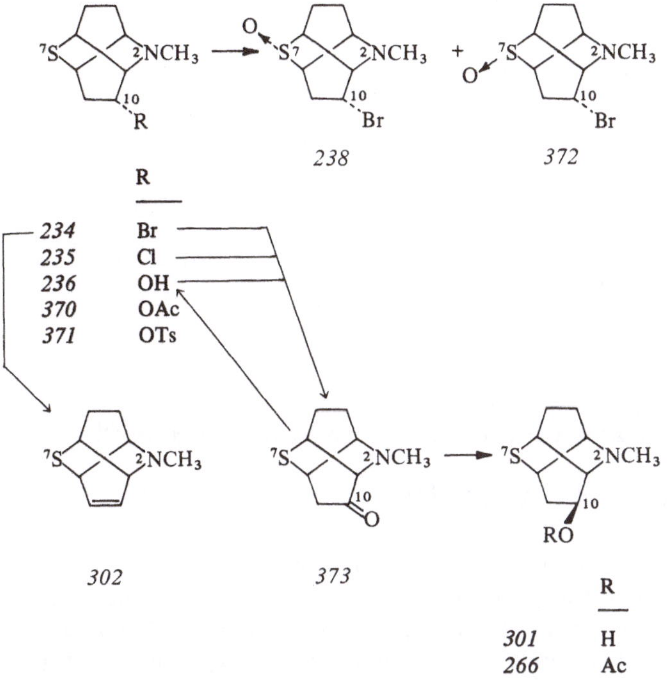

	R
234	Br
235	Cl
236	OH
370	OAc
371	OTs

302   373

	R
301	H
266	Ac

The ketone *373* was synthesized from three different starting materials. Reaction of the bromide *234* or the chloride *235* with silver tetrafluoroborate in dimethyl sulfoxide at 70° gave the ketone *373* in 79% and 75% yield, resp. In both cases also alcohol *236* was formed as by-product (8% and 20%, resp.). Oxidation of the alcohol *236* with CrO_3/pyridine gave in moderate yield the ketone *373*, too. The latter was reduced by $LiAlH_4$ to a mixture of the two C(10)-epimeric alcohols: 70% of the $10^{S(7)}$-alcohol *301* (3.2.4.) [also characterized as its acetate *266* (3.2.3.)] and 21% of the $10^{N(2)}$-alcohol *236*.

4.5. 2,8-Dihetero-homotwistbrendanes[t]

4.5.1. 2,8-Dioxa-homotwistbrendanes[50]

Oxidation of the alcohol *374*, obtained by base-hydrolysis of the acetate *311* (3.4.3.), with Jones-reagent yielded the ketone *375*. Unsubstituted 2,8-dioxa-homotwistbrendane (*377*) was prepared by converting the ketone *375* to its thioketal *376*, which on reductive desulfuration with raney-nickel gave *377*. The enolacetate *378* was available by reaction of the ketone *375* with triphenylethyllithium as base followed by addition of acetic anhydride. Under acidic conditions only decomposition could be observed and with pyridine as base no reaction took place.

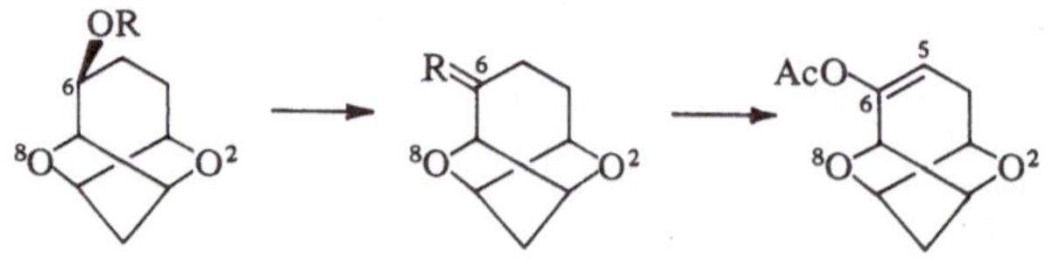

378

	R			R
311	Ac		*375*	O
374	H		*376*	S ⌐ S ⌐
			377	H$_2$

4.6. Optically Active 2,7-dioxa-isotwistane, -twistane and -twista-4,9-diene[hh)]

4.6.1. 2,7-Dioxa-isotwistane and -twistane[49, 51)]

(−)-endo-2-Hydroxy-9-oxabicyclo[3.3.1]non-6-ene (*379*) was the starting material of choice. Fractional crystallization of the diastereomeric esters (−)-*381* and (+)-*382*, obtained from the racemic alcohol (±)-*119* [(−)-*379* + (+)-*380*][48, 60, 64)] by treatment with (−)-camphanic acid chloride[82)], and subsequent LiAlH$_4$-reduction of the (−)-ester *381* gave the (−)-alcohol *379* ([α]$_D$ = −91 ± 3°). Its optical purity, checked by the ^{19}F−NMR-spectroscopic method of Dale, Dull and Mosher[83)], was ⩾ 99%. Analogous reduction of the (+)-ester *382* gave the (+)-alcohol *380* ([α]$_D$ = +80 ± 2.5°) of some less optical purity.

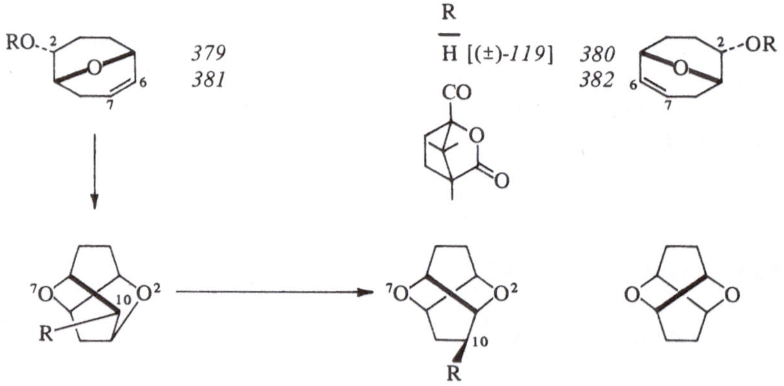

388 [(±)-*212*]

	R			R	
383	I	[(±)−*125*]	*386*	OTs	[(±)−*268*]
384	OTs	[(±)−*267*]	*387*	H	[(±)−*212*]
385	H	[(±)−*128*]			

The subsequently applied synthetic scheme was analogous to the one for the corresponding racemic compounds (see 2.2.2.1., 3.2.3. and 3.2.4.). Treatment of the (−)-alcohol *379* with iodine in chloroform yielded the (−)-10$^{O(7)}$-iodide *383* as sole product. Its reaction with silver tosylate in acetonitrile led to a mixture of 10$^{O(7)}$-tosyloxy-isotwistane *384* and -twistane *386*, which was directly treated with LiAlH$_4$ in refluxing dioxane to give a mixture, easily separable by vpc., of (−)-2,7-dioxa-isotwistane (*385*: [α]$_D$ = −23.3 ± 0.7°) and (−)-2,7-dioxa-twistane (*387*, see Table 8). The absolute configuration of *387* [(−)-(1R, 3R, 6R, 8R), right-handed helix (P)] and of all other compounds involved in its synthesis was determined by chemical correlation with (−)-(2S)-malic acid (*389*). As relais compounds served the endo-2-hydroxy-9-oxabicyclo[3.3.1]nonanes (+)-*390* and (−)-*391*, (+)-5-hydroxy-cyclooct-1-ene [(+)-*392*] and the 4-methoxy-suberic acid dimethylester (−)-*393* and (+)-*394*.

4.6.2. 2,7-Dioxa-twista-4,9-diene[25]

Treatment of racemic 4$^{O(7)}$,10$^{O(7)}$-dihydroxy-2,7-dioxa-twistane [(±)-*361* (4.4.1.): (−)-*395* + (+)-*396*] with (−)-camphanic acid chloride[82] in pyridine yielded a mixture of the two diastereomeric esters (−)-*397* and (+)-*398*, which was separated by fractional crystallization. Base-hydrolysis (1.1N K$_2$CO$_3$-solution) of (−)-*397* gave diol (−)-*395* ([α]$_D$ = −138±5°) and of (+)-*398* the diol (+)-*396* ([α]$_D$ = +132±4°), each in approx. 85−90% yield.

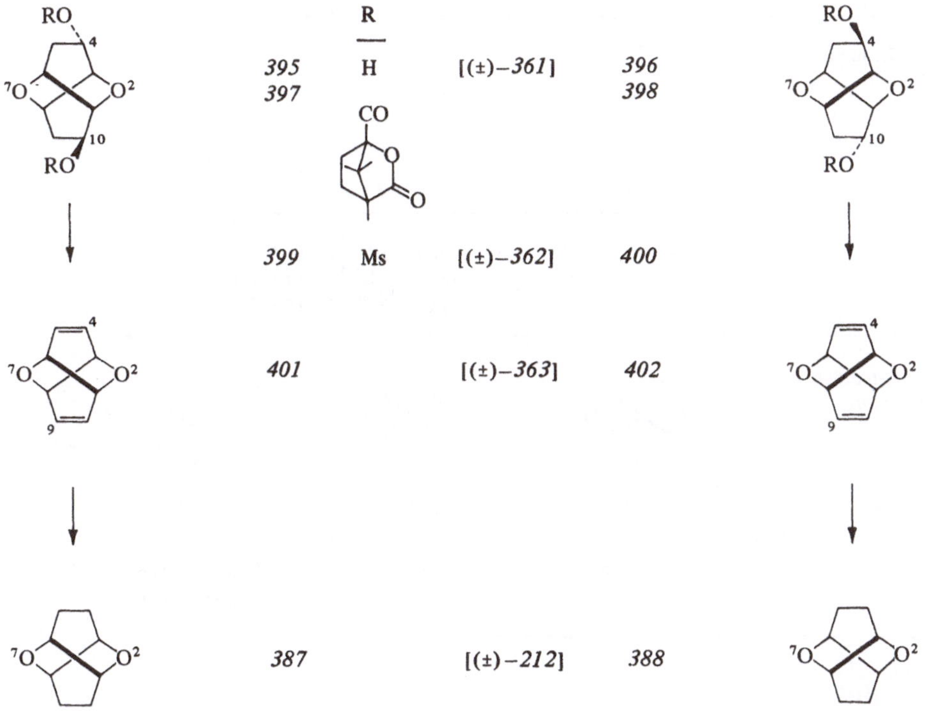

	R			
395	H	[(±)−*361*]	*396*	
397			*398*	
	CO			
399	Ms	[(±)−*362*]	*400*	
401		[(±)−*363*]	*402*	
387		[(±)−*212*]	*388*	

71

In analogy to the synthesis of racemic 2,7-dioxa-twista-4,9-diene (*363*, 4.4.1.), both enantiomeric dienes could be prepared: (−)-diol *395* → (−)-dimesylate *399* ($[\alpha]_D$ = −92.5±4°) → (−)-diene *401* (see Table 8) and (+)-diol *396* → (+)-dimesylate *400* ($[\alpha]_D$ = +93±5°) → (+)-diene *402* (see Table 8).

Catalytic hydrogenation (H_2/Pd, C) of the dienes (−)-*401* and (+)-*402* gave the corresponding saturated 2,7-dioxa-twistanes (−)-*387* (see Table 8) and (+)-*388* (see Table 8), resp. These correlations on the one hand allowed the unequivocal assignments of the absolute configuration to the dienes (−)-*401* and (+)-*402*[ii] and on the other hand give the information about their optical purity, which again is ≥ 99%.

4.6.3. Summary

On the basis of chemical correlations Tichý[84] was able to deduce the absolute configuration of carbocyclic (+)-twistane (*403*) as (1R, 3R, 6R, 8R), left-handed helix (M). Comparison of this result with the ones described in 4.6.1. and 4.6.2. demonstrate, that carbocyclic twistane and 2,7-dioxa-twistane having the same sign of the optical rotation, possess the same helicity (see Table 8).

Table 8

Right-handed helicity (P)	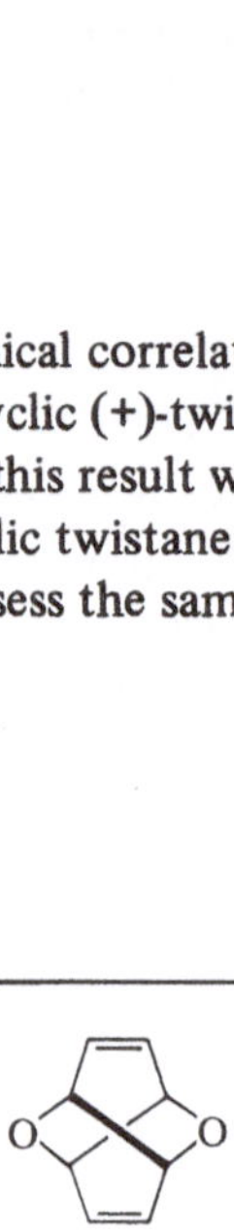 *401*	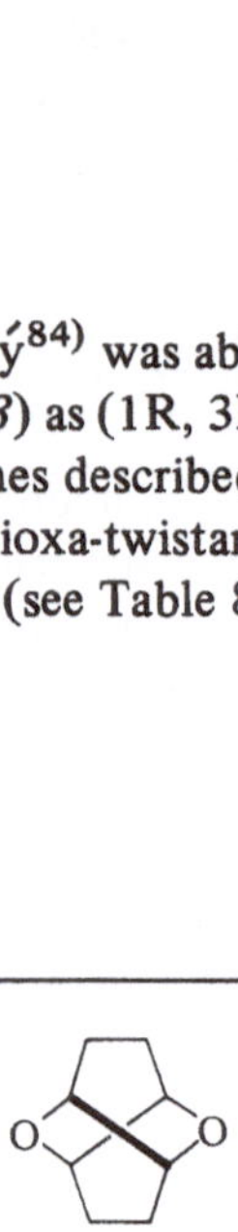 *387*	
	−321.5 ±7° [1]	−229 ± 5.5° [1] (4.6.1.) −225 ± 8° [1] (4.6.2.) −217 ± 6.5° [2] (4.6.1.)	
Left-handed helicity (M)	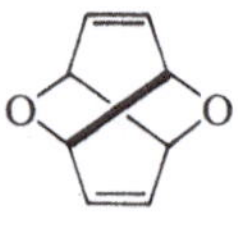 *402*	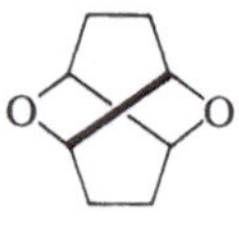 *388*	 *403*
	+326 ± 8° [1]	+222 ± 8° [1]	+434° [2] [84] +414° [2] [85]

[1] CHCl$_3$.
[2] C_2H_5OH.

5. Structural Assignments

Structural assignments (constitutions and configurations of substituents) are based on chemical conversions to known compounds and on spectroscopical measurements, mainly NMR. Not each single compound shall be discussed but rather some characteristic examples and coherences.

5.1. Chemical Correlations

Conversions within the dihetero-tricyclodecane-series on the basis of which the same or another skeleton can be assigned to the involved compounds have already been described in the Sections 2., 3. and 4.

Chemical degradations to 9-heterobicyclo[3.3.1]nonanes allowed to assign the structures to tricyclic compounds described in 5.1.1 and 5.1.2.

5.1.1. $4^{O(2)},8^{O(2)}$-Dihydroxy-2-oxa-6-thia-adamantane (*25*)[28, 48]

Desulfuration of the diol *25* (2.1.3.) with raney-nickel led to exo,exo-2,6-dihydroxy-9-oxabicyclo[3.3.1]nonane (*404*) in approx. 75% yield. Jones-oxidation of the latter

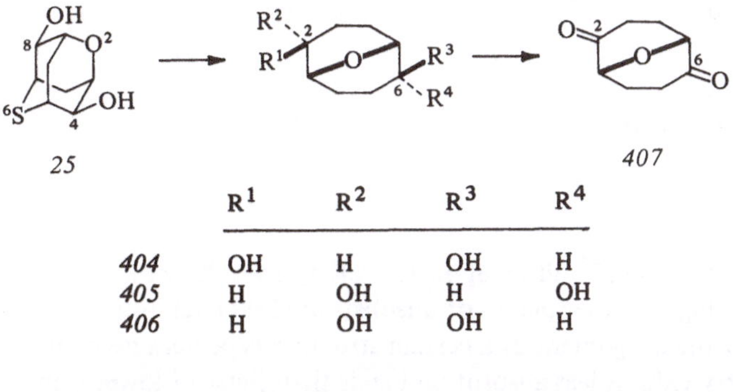

	R^1	R^2	R^3	R^4
404	OH	H	OH	H
405	H	OH	H	OH
406	H	OH	OH	H

gave the diketone *407*, which was identical with the oxidation product of the endo, endo-2,6-diol *405*. The third possible epimeric diol *406* served as further compound for comparisons.

5.1.2. 2-Oxa-7-thia-isotwistane (*148*), $10^{S(7)}$-Hydroxy-2-oxa-7-thia-isotwistane (*335*) and $10^{S(7)}$-Acetoxy-2-oxa-7-thia-twistane (*249*)

Treatment of unsubstituted 2-oxa-7-thia-isotwistane (*148*: 2.2.3.1.) with raney-nickel yielded 74% of 9-oxabicyclo[4.2.1]nonane (*408*)[86] and reductive desulfuration of the alcohol *335* (4.3.2.) 85% of endo-2-hydroxy-9-oxabicyclo[4.2.1]nonane (*409*)[60] Analogous reaction of the $10^{S(7)}$-acetoxy-twistane *249* (3.2.3.) with raney-nickel, however, led to endo-2-acetoxy-9-oxabicyclo[3.3.1]nonane (*410*)[60].

73

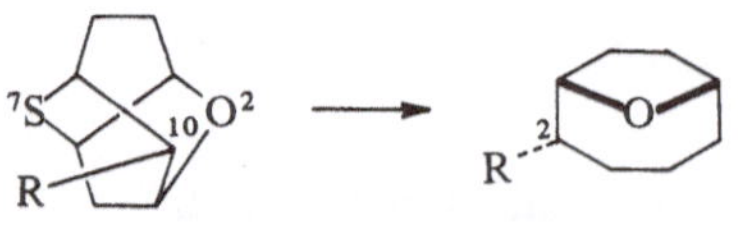

	R		R
148	H	*408*	H
335	OH	*409*	OH

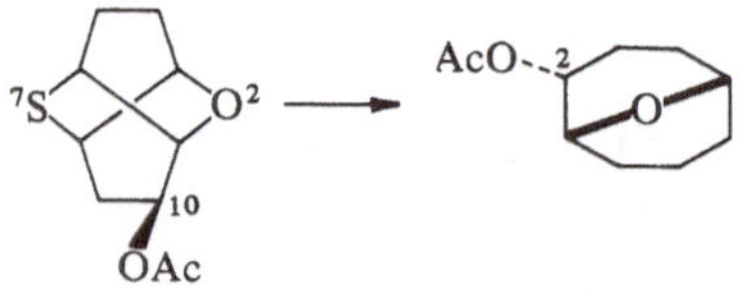

249 *410*

Together with the known structure of the bicyclic alcohol *139*, used as starting material for the above tricyclic compounds, and the position and configuration of its hydroxy-group [HO–C(2)^{endo}], the 1,4-oxygen-bridge (deduced from *408* and *409*) in the tricyclic compounds *148* and *335* and therefore in all with them chemically connected compounds, allows unequivocally the assignment of the isotwistane skeleton to all these compounds. On the other hand, the twistane skeleton for the tricyclic compound *249* follows from the 1,5-oxygen-bridge in *410* and the structure of the starting material *139*.

5.2. Spectroscopical Measurements

5.2.1. IR-Spectra

Between approx. 1200 and 800 cm^{-1} all dihetero-tricyclodecanes show marked sharp absorption bands. Comparing, *e.g.* the spectra of unsubstituted skeletal isomers, already gives good hints for the assignment of a certain structure type because compounds of higher symmetry exhibit less absorption bands than those of lower symmetry [*e.g.* 2,7-dioxa-twistane (*212*) and 2,7-dioxa-isotwistane (*128*), resp.].

Characteristic differences are also observed with skeleton-isomeric ketones of the types of isotwistane (*G 2*), twistane (*G 3*) and homotwistbrendane (*G 4*) (see Table 9). For a given combination of the two heteroatoms X and Y, twistanones absorb at highest wave numbers.

5.2.2. UV-Spectra

5.2.2.1. $10^{O(7)}$- and $10^{O(2)}$-Iodo-2,7-dioxa-isotwistanes (125 and 127)[50]

$10^{O(7)}$-Iodo-2,7-dioxa-isotwistane (*125*) absorbs at longer wave-lengths [λ_{max} = 260 nm (ϵ = 570)] than the $10^{O(2)}$-iodide *127* [λ_{max} = 257 nm (ϵ = 515)], which

74

Table 9

Y	X	G2	G3 [t]	G4 [t]	Refs.
O	O	331: 1735[1]	356: 1750[1]	375: 1720[1]	50)
S	O	336: 1725[1]	365: 1747[1]	–	53)
O	S	341: 1721[1]	–	–	27)
S	S	344: 1703/1695[1,3]	–	–	57)
S	NCH$_3$	350: 1708[2]	373[t]: 1727[2]	–	58)

[1]) In CHCl$_3$.
[2]) In CCl$_4$.
[3]) Double absorption band.

can be the effect of a stronger interaction of the $10^{O(7)}$-iodine with O(7) than of the $10^{O(2)}$-iodine with O(2). Such differences in interactions were already observed in 9-oxabicyclononanyl-iodides, where the iodines are in anti (endo-configuration) or syn (exo-configuration) position to the bridge-oxygen[60, 64, 71]. Exo-Iodines

	R^1	R^2
125	I	H
127	H	I

[syn to O(9)] exhibit a quite remarkable interaction with O(9). This is evidenced in absorption maxima at longer wave-lengths, corresponding to less energetic n → σ* transitions of the iodo-compounds. The data are summarized in Table 10.

5.2.2.2. $10^{S(7)}$- and $10^{O(2)}$-Iodides of 2-Oxa-7-thia-isotwistane and the Corresponding Sulfoxides and Sulfones[53, 54]

Characteristic differences are observed in the UV-spectra of the sulfides 146 and 147, the sulfoxides 156, 159 and 164 as well as the sulfones 162 and 163 according to the orientation of the iodines, either syn to O(2) [anti to S(7)] or syn to S(7) [anti to O(2)]. The extinction coefficient of the iodine absorption (λ_{max} approx. 260 nm) in $10^{S(7)}$-iodides is approx. 2.5-times bigger than the one in the corresponding $10^{O(2)}$-iodides (see Table 11). Furthermore, the ε-values decrease in the $10^{S(7)}$- as well as in the $10^{O(2)}$-iodides by the sequence: $\epsilon_{sulfide} > \epsilon_{sulfoxide} > \epsilon_{sulfone}$. Moreover, the absolute values partly differ strongly from those of other iodo-compounds[ii].

75

Table 10

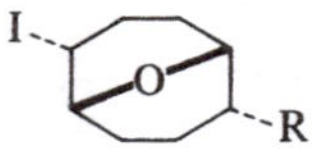

Compound	R	$\lambda_{max}(\epsilon)$
411	I	258.5 (1300)
412	H	256 (595)
413	OH	256 (620)
414	OAc	256.5 (625)

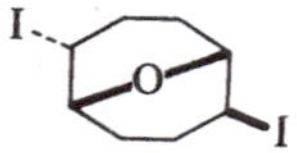

Compound	$\lambda_{max}(\epsilon)$
415	258.5 (1340)

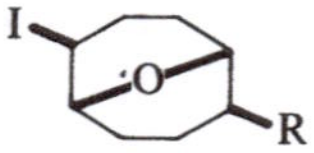

Compound	R	$\lambda_{max}(\epsilon)$
416	I	263 (1295)
417	H	263.5 (635)

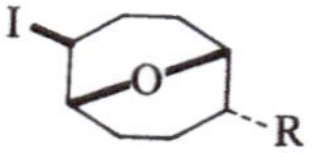

Compound	R	$\lambda_{max}(\epsilon)$
418	OH	263 (630)
419	OAc	164 (612)

Compound	$\lambda_{max}(\epsilon)$
420	256 (600)

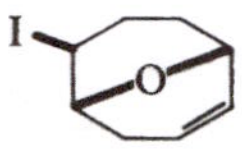

Compound	R	$\lambda_{max}(\epsilon)$
422	I	256 (1180)
423	H	258 (615)
424	OAc	257 (630)

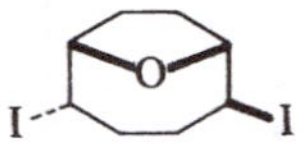

Compound	$\lambda_{max}(\epsilon)$
425	258.5 (1260)

Compound	$\lambda_{max}(\epsilon)$
421	263 (600)

Compound	R	$\lambda_{max}(\epsilon)$
426	I	263.5 (1205)
427	H	262 (595)

77

R^1	R^2				
H	I	146	156	159	162
I	H	147	–	164	163

It is interesting to note that in the sulfoxides *156* and *159*, but not in the one of *164*, absorption maxima occur (*156*: 229 nm, *159*: 239 nm), which must be assigned to transitions of the unshared electron pair at the sulfur atom. Absorption maxima at such short wave-lengths in S-oxides were among others also observed in β-keto-sulfoxides[88] and α, β-unsaturated sulfoxides[89]. They are interpreted as strong electronic interactions between a carbonyl group or a double bond with the sulfoxide group. In accordance with that the UV-data of *156* and *159* lead to the conclusion that between the iodine and the sulfoxide group such interactions can also arise. They seem, as shown by the different behaviour of *164*, to be strongly dependent on the relative spatial arrangement of the iodine and the sulfoxide group.

Table 11

	Compound	$\lambda_{max}(\epsilon)$ I–C(10)$^{O(2)}$	Compound	$\lambda_{max}(\epsilon)$ I–C(10)$^{S(7)}$
S(7)	146	264 (1820)	147	259 (744)
O–S(7)$^{C(1)}$ bb)	156	263 (sh)[1]) (1200), 229 (3750)	–	
O–S(7)$^{C(4)}$ bb)	159	258 (sh)[1]) (1115), 239 (1385)	164	263 (490)
O$_2$S(7)	162	266 (610)	163	264 (246)

[1]) As the iodine absorption maxima of *156* and *159* are shoulders (sh), the noted maxima (263 and 258 nm, resp.) may deviate somewhat from the actual accurate values.

5.2.2.3. Isotwistan-10-ones of the Types of 2-Oxa-7-thia (336), 2-Thia-7-oxa (341) and 2,7-Dithia (344) as Well as 2-Oxa-7-thia-twistan-10-one (365)[27, 53]

Keto-sulfides with the sulfur atom and the carbonyl group being separated by two saturated carbon atoms (γ-keto-sulfides) show UV-spectra which in principle correspond to the sum of the spectra of the two isolated chromophores alkyl-sulfide [λ_{max} approx. 230–240 nm (ε approx. 100–200)] and carbonyl group [λ_{max} approx. 275–285 nm (ε approx. 10–40)][90]. However, β-keto-sulfides exhibit two characteristic absorption maxima at approx. 245 nm and approx. 300 nm. The ε-values of the latter are in the order of 200–300[56, 90–93]. The UV-maxima (see Table 12) of the isotwistanones *336* and *341* therefore are characteristic for β-keto-sulfides, the ones of the twistanone *365*, however, are in good agreement for a cyclic γ-keto-sulfide.

78

Table 12

336 [53]	*341* [27]	*344* [27]	*365* [53]

$\lambda_{max}(\epsilon)$	253 (340)	262.5 (670)	254 (385)	245 (375)
	317 (250)	311 (385)	279 (590)	297 (30)
			309 (sh) (420)	
			318 (435)	

2,7-Dithia-isotwistan-10-one (*344*) actually represent a combination of the two β-keto-sulfides *336* and *341*. Its UV-spectrum indeed reflects this situation quite well exhibiting four absorption maxima.

5.2.3. NMR-Spectra

Of the various spectroscopical methods NMR-spectroscopy was the one most strongly applied and by consequence the most informative.

5.2.3.1. 2,6-Dihetero-adamantanes[kk]

5.2.3.1.1. Unsubstituted 2,6-Dihetero-adamantanes. In C-unsubstituted 2,6-dihetero-adamantanes (*G 1*) the four equivalent methylene groups H_2−C(4), H_2−C(8), H_2−C(9) and H_2−C(10) exhibit a single signal if X(2) = Y(6) but an AB-system if X(2) ≠ Y(6). The chemical shifts whose exact values are available are listed in Table 13.

In the 2-oxa-6-aza-adamantanes *58* [H−N(6)], *60* [$H_5C_2O_2$C−N(6)], *40* [H_3C−N(6)], *57* [$H_5C_6O_2$S−N(6)] and *59* [HCO−N(6)] the signals of the hydrogen atoms orientated towards Y(6) = NR [H−C(4)$^{N(6)}$, H−C(8)$^{N(6)}$, H−C(9)$^{N(6)}$ and H−C(10)$^{N(6)}$] all appear at higher field than the ones orientated towards X(2) = O [H−C(4)$^{O(2)}$, H−C(8)$^{O(2)}$, H−C(9)$^{O(2)}$ and H−C(10)$^{O(2)}$]: $\Delta\nu_{AB}$ = 0.1−0.25 ppm. Just the opposite situation is observed in 2-thia-6-oxa-adamantane (*28*, see footnote [2]) in Table 13) and in 2-thia-6-aza-adamantanes *100* [H−N(6)], *99* [$H_5C_2O_2$C−N(6)] and *98* [H_3C−N(6)]. In these compounds the signals of the hydrogen atoms orientated towards Y(6) = NR are all at lower field than those orientated toward X(2) = S. Some shifts are quite remarkable. In general, by replacing O(2) by S(2) the signal for H−C(4)$^{Y(6)}$ [or H−C(8)$^{Y(6)}$ or H−C(9)$^{Y(6)}$ or H−C(10)$^{Y(6)}$, resp.] is clearly more shifted ($\Delta\delta_{Y(6)}$) to lower a field than the one ($\Delta\delta_{X(2)}$) for H−C(4)$^{X(2)}$ [or H−C(8)$^{X(2)}$ or H−C(9)$^{X(2)}$ or H−C(10)$^{X(2)}$, resp.]. This is especially the case in 6-aza-adamantanes, which show an almost constant value of −0.50 to −0.52 ppm for $\Delta\delta_{Y(6)}$.

Table 13

Compound	Y(6)	X(2)	δH–C(4)X(2)[1]	$\Delta\delta$ X(2)	δH–C(4)Y(6)[1]	$\Delta\delta$ Y(6)	Refs.
14	O	O	2.00		2.00		20, 21, 24, 26)
28[2]	O	S	2.26	-0.26	2.39	-0.39	28)
58	NH	O	2.09		1.89		36)
100	NH	S	2.31	-0.22	2.41	-0.52	40)
60	$NCO_2C_2H_5$	O	2.08		1.83		36)
99	$NCO_2C_2H_5$	S	2.21	-0.13	2.35	-0.52	40)
40	NCH_3	O	2.04		1.93		36)
98	NCH_3	S	2.19	-0.15	2.43	-0.50	40)
57	$NSO_2C_6H_5$	O	1.94		1.70		36)
59	NCHO	O	1.8		2.15 and 2.2[3]		36)
104	NH	NH	1.98		1.98		32)

[1]) Or H–C(8) or H–C(9) or H–C(10), resp.
[2]) For better comparisons of the data the normal notation O(2), S(6) has been changed to S(2), O(6).
[3]) 1 : 1-Mixture of the possible rotamers due to hindered rotation of the amide bond (C–N) (see *e.g.* 95, 96)) giving rise to two AB-systems. The B-parts at higher field fall together (δ = 1.8) and correspond to H–C(4)O(2), H–C(8)O(2), H–C(9)O(2) and H–C(10)O(2).

5.2.3.1.2. 4,8-Disubstituted 2,6-Dihetero-adamantanes[ll]. The methylene groups
$H_2-C(9)$ and $H_2-C(10)$ in 4,8-disubstituted 2,6-dihetero-adamantanes appear generally independent of the nature of X and Y

as a single AB-system for	$G\ 83$: $R^1 = R^2$
	$G\ 85$: $R^1 = R^2$
as two AB-systems for	$G\ 83$: $R^1 \neq R^2$
	$G\ 84$: $R^1 \neq R^2$
	$R^1 = R^2$
	$G\ 85$: $R^1 \neq R^2$

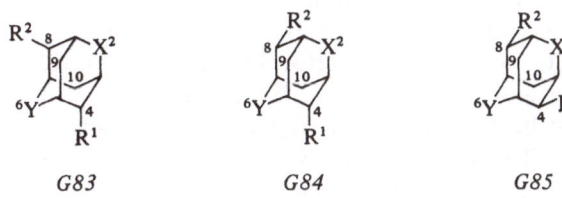

| $G83$ | $G84$ | $G85$ |

In this section several characteristic NMR-data of 2-oxa-6-aza-adamantanes
[X(2) = O, Y(6) = NR] shall be discussed standing as typical examples for the rather
large number of 4,8-disubstituted 2,6-dihetero-adamantanes known so far.

In all the studied compounds of the type $G\ 83$ [X(2) = O, Y(6) = $NSO_2C_6H_5$:
47–49) or $G\ 85$ [X(2) = O, Y(6) = $NSO_2C_6H_5$: 54–56) in each case at lower
field appearing A-part of the AB-system belongs to the H-atoms, which are in 1,3-
diaxial position to the substituents at C(4) and C(8), *i.e.* $H-C(9)^{Y(6)}$ and
$H-C(10)^{Y(6)}$ in $G\ 83$ or $H-C(9)^{X(2)}$ and $H-C(10)^{X(2)}$ in $G\ 85$, resp, The H-atoms
$H-C(9)^{X(2)}$ and $H-C(10)^{X(2)}$ in $G\ 83$ or $H-C(9)^{Y(6)}$ and $H-C(10)^{Y(6)}$ in $G\ 85$ belong to the at higher field appearing B-part. In these cases the A-part, compared to

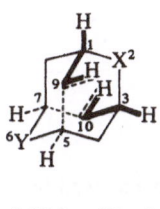

G 86 $J_{1,9}X > J_{5,9}X$
$J_{3,10}X > J_{7,10}X$

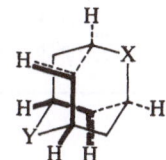

G 87 $J_{1,9}Y < J_{5,9}Y$
$J_{3,10}Y < J_{7,10}Y$

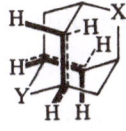

G 88 $J_{1,9}\ X > J_{1,9}Y$
$J_{3,10}X > J_{3,10}Y$

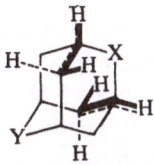

G 89 $J_{5,9}\ X < J_{5,9}Y$
$J_{7,10}X < J_{7,10}\ Y$

the B-part, is always a signal with a smaller bandwidth at half height ($w^{1}/_{2}$) and with a more distinctive line pattern. On the basis of spin, spin-decoupling experiments it was possible to determine the particular vicinal coupling constants (J_{vic}) between the methylene H-atoms at C(9) and C(10) on the one hand and the bridgehead H-atoms at C(1), C(3), C(5) and C(7) on the other hand, see G 86–G 89[mm]. Although the dihedral angles between the bridgehead and methylene H-atoms (on the basis of the symmetry of the adamantane skeleton) are all approx. 60°, in principle two different groups of vicinal coupling constants can be observed: one with J approx. 1.5–2.5 Hz (characterized by the H-atoms connected through – – – – lines) and another with J approx. 3.5–4.5 Hz (characterized by the H-atoms connected by ——— lines). This reflects the influence of the heteroatoms X(2) [or Y(6), resp.] to reduce J_{vic} between a bridgehead H-atom geminal and a methylene H-atom vicinal to X(2) [or Y(6), resp.]. The effect becomes larger if X(2) [or Y(6), resp.] is arranged trans-coplanar to the methylene H-atom (approx. 1.5–2.5 Hz) and smaller if the angle between X(2) [or Y(6), resp.] and the methylene H-atom is approx. 60° (approx. 3.5–4.5 Hz)[nn]. The vicinal coupling constants of diacetate 80 as an illustrative example are listed in Table 14.

Table 14[36)]

80

Group 1.5 – 2.5 Hz	Group 3.5 – 4.5 Hz
$J_{1,9}N(6) = J_{3,10}N(6) = 2$	$J_{1,9}O(2) = J_{3,10}O(2) = 4.5$
$J_{1,8}N(6) = J_{3,4}N(6) = 1.5$	$J_4N(6)_5 = J_{7,8}N(6) = 4$
$J_{5,9}O(2) = J_{7,10}O(2) = 2$	$J_{5,9}N(6) = J_{7,10}N(6) = 3.5$

From the data in Table 14 it further can be seen that the same is also caused by the electronegative atoms O(2) and N(6) on the vicinal coupling constants between the bridgehead H-atoms and the corresponding H-atoms H–C(4)[N(6)] and H–C(8)[N(6)] geminal to an acetoxy group, independent of the latter substituents.

Substitution of one or both iodine atoms by acetoxy groups in the 4,8-diiodo-2-oxa-6-aza-adamantanes 47 (G 83), 54 (G 85) and 50 (G 84) effects characteristic changes of the positions of the signals corresponding to the methylene H-atoms at C(9) and C(10). The results are summarized in the Tables 15–17.

Table 15: in 4[N(6)],8[N(6)]-disubstituted compounds (type G 83) the signals for H–C(10)[N(6)] or H–C(9)[N(6)], resp., are shifted to higher field by 0.39 or 0.43 ppm, resp., if the iodine atoms fixed in a 1,3-diaxial position to these H-atoms are replaced by acetoxy groups (47 → 48 or 48 → 49), whereas the positions of the signals for

Table 15[35, 36)]

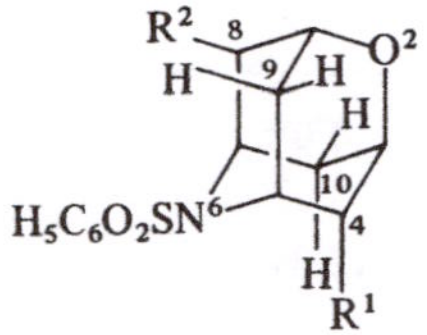

	R¹	R²	δH–C(9)O(2)	ΔδH–C(9)O(2)	δH–C(9)N(6)	ΔδH–C(9)N(6)	δH–C(10)O(2)	ΔδH–C(10)O(2)	δH–C(10)N(6)	ΔδH–C(10)N(6)
47	I	I	2.22		2.97		2.22		2.97	
48	OAc	I	2.09	+0.13	2.80	+0.17	2.18	+0.04	2.58	+0.39
48	OAc	I	2.09		2.80		2.18		2.58	
49	OAc	OAc	2.03	+0.06	2.37	+0.43	2.03	+0.15	2.37	+0.21

Table 16[35, 36)]

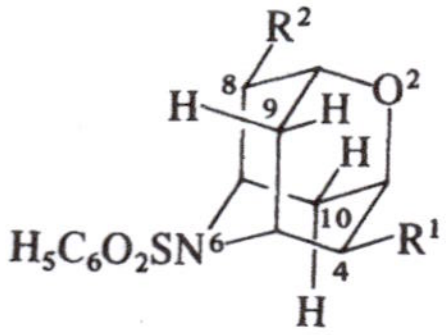

	R^1	R^2	δH–C(9)O(2)	$\Delta\delta$H–C(9)O(2)	δH–C(9)N(6)	$\Delta\delta$H–C(9)N(6)	δH–C(10)O(2)	$\Delta\delta$H–C(10)O(2)	δH–C(10)N(6)	$\Delta\delta$H–C(10)N(6)
54	I	I	3.04	+0.57	1.92	+0.35	3.04	+0.03	1.92	−0.07
55	OAc	I	2.47		1.57		3.01		1.99	
55	OAc	I	2.47	+0.04	1.57	−0.07	3.01	+0.58	1.99	+0.35
56	OAc	OAc	2.43		1.64		2.43		1.64	

Table 17[00] 35, 36)

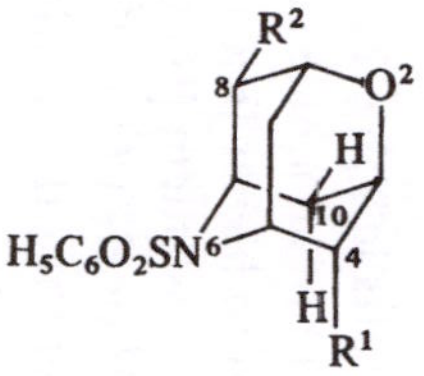

	R¹	R²	$\delta_{H-C(10)O(2)}$	$\Delta\delta_{H-C(10)O(2)}$	$\delta_{H-C(10)N(6)}$	$\Delta\delta_{H-C(10)N(6)}$
50	I	I	2.88		2.53	
51	I	OAc	2.31	+0.57	2.54	–0.01
50	I	I	2.88		2.53	
52	OAc	I	2.87	+0.01	1.93	+0.60

Table 18[35, 36)]

$$R^4 \underset{8}{\diagdown} R^3 \diagdown O^2$$

H$_5$C$_6$O$_2$SN–$\underset{4}{\overset{6}{\diagdown}}$–R^1

R^2

Compound	R^1	R^2	R^3	R^4	δ H–C(4)O(2)	δ H–C(4)N(6)
a. 73	H	OAc	H	H	4.91	–
52	H	OAc	I	H	4.89	–
48	H	OAc	H	I	4.85	–
49	H	OAc	H	OAc	4.85	–
b. 56	OAc	H	OAc	H	–	4.64
55	OAc	H	I	H	–	4.4–4.65
51	OAc	H	H	I	–	4.65
c. 50	H	I	I	H	4.64	–
51	H	I	OAc	H	4.65	–
47	H	I	H	I	4.79	–
48	H	I	H	OAc	4.57	–
d. 54	I	H	I	H	–	4.38
55	I	H	OAc	H	–	4.4–4.65
50	I	H	H	I	–	4.64
52	I	H	H	OAc	–	4.44

H–C(10)$^{O(2)}$ or H–C(9)$^{O(2)}$, resp., remain almost unchanged (shifts: 0.04 or 0.06 ppm). It is interesting to note that by replacing 4$^{N(6)}$-I by 4$^{N(6)}$-OAc also the methylene H-atoms H–C(9)$^{O(2)}$ and H–C(9)$^{N(6)}$ (0.13 and 0.17 ppm) and by replacing 8$^{N(6)}$-I by 8$^{N(6)}$-OAc the H–C(10)$^{O(2)}$ and H–C(10)$^{N(6)}$ (0.15 and 0.21 ppm) are shifted to higher field.

Table 16: in contrast to the results in Table 15, in 4$^{O(2)}$,8$^{O(2)}$-disubstituted compounds (type G 85) replacement of an iodine atom by an acetoxy group (54 → 55 or 55 → 56) effects a shift to higher field not only of the methylene H-atom in 1,3-diaxial position to the substituent [H–C(9)$^{O(2)}$ or H–C(10)$^{O(2)}$], but also of its geminal one [H–C(9)$^{N(6)}$ or H–C(10)$^{N(6)}$] (0.57 or 0.58 ppm and 0.35 or 0.35 ppm, resp.). The H-atoms of the other methylene groups, however, suffer no remarkable shifts (0.03–0.07 ppm).

Table 17: in 4$^{N(6)}$,8$^{O(2)}$-disubstituted compounds (type G 84) replacement of an iodine atom by an acetoxy group (50 → 51 or 50 → 52) shifts the methylene H-atom H–C(10)$^{O(2)}$ [or H–C(10)$^{N(6)}$] in 1,3-diaxial position to R^2 (or to R^1, resp.) by approx. 0.6 ppm to higher field, whereas the other one H–C(10)$^{N(6)}$ [or H–C(10)$^{O(2)}$] suffers practically no change. In this case the influence of O(2) as well as of the sulfonamide group H$_5$C$_6$O$_2$S–N(6) on H$_2$–C(10) must either be very small or approx. in the same order[pp)].

The chemical shifts of the H-atoms at C(4) [or C(8), resp.], which are geminal to an acetoxy group are neither substantially influenced by the nature nor the configuration of the substituent at the opposite carbon C(8) [or C(4), resp.], see Table 18, parts a and b. However, H-atoms geminal to an iodine atom are sometimes shifted quite remarkably, see Table 18, parts c and d.

From the data in Table 18, parts a and b, it can also be seen, that the configuration of an acetoxy group at C(4) [or C(8), resp.] can be deduced on the basis of the chemical shift of the H-atom geminal to it.

The signal of H–C(4)$^{O(2)}$ [or H–C(8)$^{O(2)}$, resp.] in $4^{N(6)},8^{N(6)}$-diiodide 47 (see Table 18, part c) is clearly shifted to lower field (deshielding effect) than in the other compounds. This might be caused by a hindered rotation of the sulfonamide bond (N–S) resulting from the large iodine atoms orientated toward the sulfone group.

5.2.3.2. 2,7-Dihetero-isotwistanes and -twistanes

5.2.3.2.1. Unsubstituted 2,7-Dihetero-isotwistanes and -twistanes

5.2.3.2.1.1. General Aspects. Whereas in unsubstituted isotwistanes (*G 2*) with X(2) $\neq$ Y(7) as well as with X(2) = Y(7) (C_1-symmetry) the signals of the four bridgehead H-atoms H–C(1), H–C(3), H–C(6) and H–C(8) a priori may have different chemical shifts (see Table 19), the same H-atoms in unsubstituted twistanes (*G 3*) effect a single signal with X(2) = Y(7) (D_2-symmetry) or two signals with X(2) $\neq$ Y(7) (C_2-symmetry) (see Table 20). By consequence the number of signals is a valuable standard for the differentiation between unsubstituted 2,7-dihetero-isotwistanes (*G 2*) and -twistanes (*G 3*).

Furthermore, in isotwistanes (*G 2*) of the two H-atoms H–C(1) and H–C(3) on one the hand and H–C(6) and H–C(8) on the other hand (in any case on the ones or the others of the bridgeheads adjacent to a distinct heteroatom), in each case the former one, *i.e.* H–C(1) and H–C(6), resp., appears at higher field than the other one, *i.e.* H–C(3) and H–C(8), resp.

5.2.3.2.1.2. 2,7-Dioxa-twistane (212)[50], -twist-4-ene (360)[61] and -twista-4,9-diene (363)[61]. The NMR-spectrum of 2,7-dioxa-twistane (*212*), which consists of two multiplets at δ = 1.5–2.3 ppm and δ = 3.85 ppm in the ratio of 2 : 1 corresponds to the general type of an AA'A''A'''BB'B''B'''XX'X''X'''-spectrum. The great difference between the chemical shifts of the methylene H-atoms (AA'A''A'''BB'B''B''') and the ones at the bridgeheads (XX'X''X''') enable the simplification of the complex spectrum by spin, spin-decoupling experiments to one of a typical AA'BB'-type, symmetrical with the centre at δ = 1.9 ppm (the spectra were recorded at 60 and 100 MHz). With the aid of the computer program LAOCOON III, which is a modified variation of the program LAOCOON II[97], the chemical shift difference and the coupling constants of the AA'BB'-part[qq] were determined (see Table 21). The parameters of the observed and calculated spectra were in good agreement[rr].

The spectrum of 2,7-dioxa-twist-4-ene (*360*) shows the following signals: 1.7–2.4/m, H_2–C(9), H_2–C(10) [typical AA'BB'-spectrum on simultaneous irradiation

Table 19. Isotwistanes (*G 2*)

Compound	Y(7)	X(2)	Chemical shifts (δ[ppm])				Refs.
			H–C(1)	H–C(3)	H–C(6)	H–C(8)	
128	O	O	4.1–4.4 (3H)[1], 4.51 (1H)[1]				50)
148	S	O	4.46	4.65	3.16	3.34	53)
165	SO$_2$	O	4.46	4.79	3.29	3.50	54)
171	NCO$_2$C$_2$H$_5$	O	4.25–4.75 (4H)[1]				55)
172	NCH$_3$	O	4.1–4.3 (2H)		2.84	3.18	55)
173	NH	O	4.2–4.5 (2H)		3.2–3.6 (2H)		55)
182	O	S	3.52	3.90	4.20	4.56	27)
200	S	S	3.45–3.8	4.17	3.22	3.45–3.8	57)
295 · HCl	NCH$_3$ · HCl	S	3.4–3.75 (2H)		4.0–4.3 (2H)		58)
297 · HCl	O	NCH$_3$ · HCl	3.6–4.05 (2H)		4.2–4.65 (2H)		58)
299 · HCl	S	NCH$_3$ · HCl	3.34 (1H)[1], 3.5–3.95 (2H)[1] and 4.12 (1H)[1]				58)

[1]) Not assigned.

Table 20. twistanes (*G 3*)

Compounds	Y(7)	X(2)	Chemical shifts (δ[ppm])		Refs.
			H–C(1) and H–C(3)	H–C(6) and H–C(8)	
212	O	O	3.85	3.85	50)
289	S	O	4.06	2.95	53)
298 · HCl	NCH$_3$ · HCl	O	3.3–3.85	3.95–4.25	58)
296 · HCl^t)	S	NCH$_3$ · HCl	3.5–4.2	3.2–3.5	58)

Table 21

Chemical shift difference	Coupling constants [Hz]			
[Hz] $\Delta\nu_{AB}$	$J_9O(2),_9O(7)$ $J_{10}O(2),_{10}O(7)$	$J_9O(2),_{10}O(2)$ $J_9O(7),_{10}O(7)$	$J_9O(2),_{10}O(7)$	$J_9O(7),_{10}O(2)$
33.2 ± 0.4	−13.5 ± 0.3	+8.5 ± 0.2	+7.3 ± 0.3	+0.9 ± 0.2

of the H−C(1) and H−C(8) nuclei in a double resonance experiment]; 3.64/d, $J_{1,10} = J_{8,9} = 2.4$ Hz (further splitting by $J_{1,6} = J_{3,8} = 1.7$ Hz and $J_{1,3} = J_{6,8} = 0.6$ Hz), H−C(1) and H−C(8); 4.42/q, $J_{1,6} = J_{3,8} = 1.7$ Hz, $J_{3,4} = J_{5,6} = 1.7$ Hz, $J_{3,5} = J_{4,6} = 1.7$ Hz (further splitting by $J_{1,3} = J_{6,8} = 0.6$ Hz), H−C(3) and H−C(6); 6.70/t, $J_{3,4} = J_{5,6} = 1.7$ Hz, $J_{3,5} = J_{4,6} = 1.7$ Hz, H−C(4) and H−C(5).

The NMR-spectrum of 2,7-dioxa-twista-4,9-diene (363) consists of two multiplets at 4.24 [$w^{1}/_2$ approx. 6 Hz, H−C(1), H−C(3), H−C(6) and H−C(8)] and 6.61 [$w^{1}/_2$ approx. 6 Hz, H−C(4), H−C(5), H−C(9) and H−C(10)]. Both signals are changed to singlets in double irradiation experiments.

5.2.3.2.2. C(10)-Substituted 2,7-Dihetero-isotwistanes and -twistanes[t)]

5.2.3.2.2.1. 2,7-Dihetero-isotwistanes in General. NMR-data are a proper tool for the assignment of the configuration of the substituents at C(10) in 2,7-dihetero-iso-twistanes and -twistanes[t)]. A not further split doublet with a coupling constant of approx. 11−12 Hz at δ approx. ⩾ 2.0 ppm is characteristic for a $C(10)^{Y(7)}$-substituted isotwistane G 90. This signal of a single methylene H-atom clearly at lower field than the remaining ones belongs to $H-C(9)^{Y(7)}$. The substituent at C(10) is arranged syn to Y(7) [anti to X(2), resp.] and therefore in a quasi 1,3 diaxial relationship to $H-C(9)^{Y(7)}$, which explains the strong deshielding effect on the latter. The coupling constant of 11−12 Hz corresponds to the geminal coupling between $H-C(9)^{Y(7)}$ and $H-C(9)^{X(2)}$, $J_{1,9}^{Y(7)}$ and $J_{8,9}^{Y(7)}$ both being practically zero[ss)]. For an illustration corresponding nmr.-data of $10^{O(7)}$-substituted 2,7-dioxa-isotwistanes (G 90) are listed in Table 22.

G90   G91

On the basis of model studies one could·expect that also in $C(10)^{X(2)}$-substituted isotwistanes (G 91) a single methylene H-atom, namely $H-C(5)^{X(2)}$, will be strongly deshielded and because of the special arrangement of its vicinal H-atoms $H_2-C(4)$ and H−C(6), however, should show a more complex coupling pattern than only a

Table 22

Compound G 90 [X(2) = O, Y(7) = O]	$R-C(10)^{O(7)}$	$\delta H-C(9)^{O(7)}$ [ppm]	$J_9O(2), {}_9O(7)$ [Hz]	Refs.
120	HgOAc	2.08	12	50)
121	HgBr	2.10	12	50)
122	HgI	2.16	11	50)
123	HgNO$_3$	$-^1$)	$-^1$)	50)
124	HgCl	1.99	11.5	50)
125	I	2.84	12	50)
126	Br	2.77	12	50)
137	OH	2.45	12	50)
138	OTHP	2.43 and 2.46^2)	11	48)
246	OAc	2.36	12	50)
267	OTs	2.38	12	50)
287	N$_3$	2.37	12	50)
330	NH$_2$	2.22	12	50)

[1]) Compound 123 has not been isolated.
[2]) Two doublets as a result of the two diastereomeric pairs of enantiomers.

doublet. Indeed, this can be observed in the NMR-spectra of $10^{X(2)}$-substituted isotwistanes, where the deshielding effect of $R-C(10)^{X(2)}$ is big enough to separate the signal of $H-C(5)^{X(2)}$ from the multiplets of the methylene H-atoms $H_2-C(4)$ and $H_2-C(9)$. E.g. $10^{O(2)}$-iodo-2,7-dioxa-isotwistane [127 (G 91): X(2) = O, Y(7) = O, R = I] the signal of $H-C(5)^{O(2)}$ appears at $\delta = 2.55$ ppm with the geminal coupling constant $J_5{}^{O(2)},{}_5{}^{O(7)} = 14$ Hz and the vicinal ones $J_4{}^{O(2)},{}_5{}^{O(2)}$ and $J_4{}^{O(7)},{}_5{}^{O(2)}$ each of 9 Hz^{tt}).

A further possibility of deducing the configuration of a substituent at C(10) in 2,7-dihetero-isotwistanes is given by the signal of $H-C(10)$ itself. In $C(10)^{Y(7)}$-substituted compounds (G 90) the coupling constants between $H-C(10)^{X(2)}$ and the neighboring vicinal H-atoms are approx. 4–5 Hz for $J_{1,10}{}^{X(2)}$ and approx. 0.5–2 Hz for $J_{6,10}{}^{X(2)}$. However, in $C(10)^{X(2)}$-substituted isotwistanes (G 91) the coupling constants of $H-C(10)^{Y(7)}$ are approx. 1–3 Hz for $J_{1,10}{}^{Y(7)}$ and approx. 6 Hz for $J_{6,10}{}^{Y(7)}$. The data of $10^{O(7)}$-substituted 2,7-dioxa-isotwistanes [G 90: X(2) = O, Y(7) = O] are listed as examples in Table 23 and those of $10^{O(2)}$-substituted 2,7-dioxa-isotwistanes [G 91: X(2) = O, Y(7) = O] in Table 24. Analogous data for $10^{S(7)}$ and $10^{O(2)}$-substituted 2-oxa-7-thia-isotwistanes [G 90 and G 91: X(2) = O, Y(7) = S] are summarized in Table 25 (see 5.2.3.2.2.2.).

5.2.3.2.2.2. 2-Oxa-7-thia-isotwistanes as Well as Corresponding Sulfoxides and Sulfones.
Applying the results of extensive studies (using the NMR-method) of the anisotropic effects in sulfoxides by Foster et al.[98] (see also[99, 100]) the NMR-sepctra of 2-oxa-7-thia-isotwistane S(7)-oxides were interpreted and in each case the configuration of the sulfoxide group had been assigned. In Table 25 the NMR-data of comparable 2-oxa-7-thia-isotwistane-sulfides, -sulfoxides and -sulfones are listed. Each of the H-atoms $H-C(9)^{S(7)}$ and $H-C(10)^{S(7)}$ is in a quasi 1,3-syn-axial relationship to the

Table 23[50]

Compound G 90	R–C(10)$^{O(7)}$	$\delta_{H-C(10)}O(2)$ [ppm]	$J_{1,10}O(2)$ [Hz]	$J_{6,10}O(2)$ [Hz]
120	HgOAc	2.58	4	2
121	HgBr	2.72	4	2
122	HgI	2.79	4	2
124	HgCl	2.53	4	2
125	I	4.52	5	⩽ 1
137	OH	3.65	4.5	⩽ 1
246	OAc	4.63	5	⩽ 1
287	N$_3$	3.48	4.5	⩽ 1
330	NH$_2$	2.71	5	⩽ 1

Table 24[50]

Compound G 91	R–C(10)$^{O(2)}$	$\delta_{H-C(10)}O(7)$ [ppm]	$J_{1,10}O(7)$ [Hz]	$J_{6,10}O(7)$ [Hz]
310	OAc	4.86	1	6.5
333	OH	3.76	< 1	6
334	OTs	4.56	1	6

O-atom of the sulfoxide [S(7)$^{C(1)}$-oxides[bb)]] or to the orbital of the free electron pair at the sulfur atom [S(7)$^{C(4)}$-oxides[bb)]] and therefore becomes deshielded or shielded, resp. Furthermore, the relative positions of the signals corresponding to the H-atoms, which are vicinal to the sulfoxide group, i.e. H–C(6) and H–C(8) correlate characteristically to the O–S(7)-configuration: in sulfoxides with S(7)$^{C(1)}$-configuration in each case H–C(8) appears at lower field than H–C(6) (e.g. 156: $\delta_{H-C(6)}$ = 3.25 ppm, $\delta_{H-C(8)}$ = 3.90 ppm), however, in compounds with S(7)$^{C(4)}$-configuration the other way around (e.g. 159: $\delta_{H-C(6)}$ = 3.81 ppm, $\delta_{H-C(8)}$ = 3.55 ppm).

In the multiplets of H–C(8) among others, a coupling constant in the order of 2 Hz (approx. 1–2.5 Hz) is observed. Spin,spin-decoupling experiments proved unequivocally that this is the result of a long-range coupling between H–C(6) and H–C(8) ($J_{6,8}$), which is characteristic for a quasi "planar M-arrangement" of an (H–C–Z–C–H)-group (Z = C, O, S, see[101)]. It was found that this coupling constant $J_{6,8}$ has even a bigger value (approx. 3–3.5 Hz) in isotwistane-sulfones than in corresponding sulfoxides (approx. 1.5–2.5 Hz) and sulfides (approx. 1–2 Hz).

From the NMR-data of isotwistane-sulfones one sees that the H-atoms influenced by the O-atoms of the sulfone, especially H–C(9)$^{S(7)}$, H–C(10)$^{S(7)}$ as well as H–C(6) and H–C(8), are deshielded. The anisotropic effect of the sulfone group is obviously weaker and less differentiated than the one of the sulfoxide group (see also[99)]).

5.2.3.2.2.3. 2,7-Dihetero-twistanes in General. In substituted 2,7-dihetero-twistanes the position of the substituent [at C(10)$^{t)}$] and its configuration can easily be deter-

Table 25[53, 54)]

Type of compound	Compound	n	Config. at S(7)	H–C(1)	H–C(3)	H–C(6)	H–C(8)	H–C(9)S(7)	H–C(10)[1)]
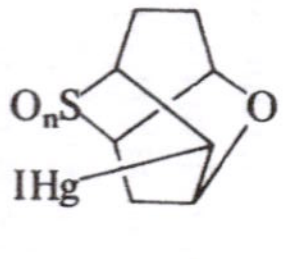 G93	*144*	0		4.7[2)]	4.7[2)]	3.78	3.36	2.50	3.09
	155	1	C(1)	4.56	4.74	3.49	3.71	3.06	2.39
	158	1	C(4)	4.60	4.79	4.40	3.55	< 2.5	3.32
	161	2		4.8[2)]	4.8[2)]	3.98	3.76	3.01	3.1[2)]
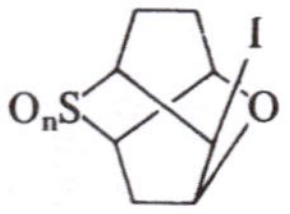 G94	*146*	0		4.7[2)]	4.7[2)]	3.28	3.52	< 2.5	5.06
	156	1	C(1)	4.66	4.76	3.25	3.90	3.24	5.20
	159	1	C(4)	4.5[2)]	4.73	3.81	3.55	1.97	4.5[2)]
	162	2		4.8[2)]	4.64	3.39	3.61	2.90	4.8[2)]
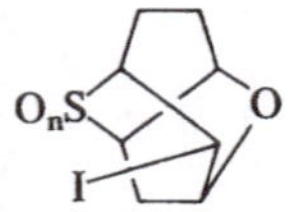 G95	*147*	0		4.5[2)]	4.72	3.25[2)]	3.25[2)]	3.05	4.5[2)]
	164	1	C(4)	4.4[2)]	4.80	3.92	3.47	2.81	4.4[2)]
	163	2		4.4[2)]	4.81	3.5[2)]	3.5[2)]	3.42	4.4[2)]

	n							
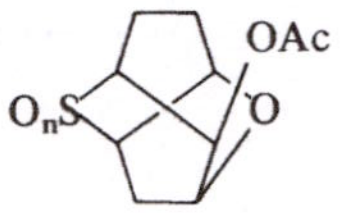 *312*	0		4.32	4.63	$3.3^{2)}$	3.3	<2.4	5.08
339	1	C(1)	4.41	4.71	3.38	3.97	3.15	5.32
G96								
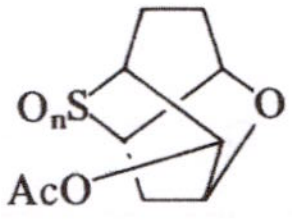 *248*	0		4.39	4.62	3.03	3.30	2.63	4.85
338	1	C(1)	4.45	4.73	3.23	3.85	3.51	4.98
250	1	C(4)	4.29	4.71	$3.6^{2)}$	$3.6^{2)}$	<2.5	5.17
252	2		4.38	4.81	3.38	3.54	3.24	5.09
G97								
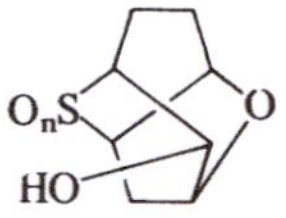 *335*	0		4.36	4.85	2.95	3.22	2.56	3.58
337	1	C(1)	4.55	4.75	3.36	3.96	3.53	$4.05^{2)}$
$428^{3)}$	2		4.40	4.97	$3.3^{2)}$	3.55	3.13	4.19
G98								

[1] H–C(10)O(2) or H–C(10)S(7) in each case corresponding to the substituent at C(10).
[2] Exact values could not be determined.
[3] Prepared from *250*[54].

Table 26

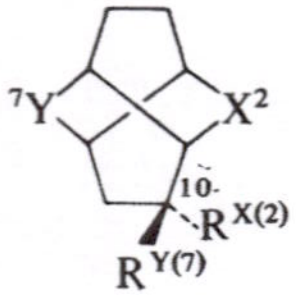

| | | | | | Approx. coupling constants [Hz] | | | | | | |
Compound	Y(7)	X(2)	$R^{Y(7)}$	$R^{X(2)}$	$J_9X(2),9Y(7)$	$J_{1,10}X(2)$	$J_9X(2),10X(2)$	$J_9Y(7),10X(2)$	$J_9X(2),10Y(7)$	$J_9Y(7),10Y(7)$	Refs.
212	O	O	H	H	13.5	1)	8.5	0.9	7.3	8.5	50)
247			OAc	H	1)	4	6.5	2	–	–	50)
268			OTs	H	1)	5	8	1	–	–	50)
355			NH$_2$	H	1)	5	8.5	1	–	–	50)
359			H	NH$_2$	12.5	–	–	–	6	8.5	50)
249	S	O	OAc	H	14	6	8	1	–	–	53)
270			OTs	H	14	6	7.5	1	–	–	53)
364			OH	H	14	1)	7.5	1)	–	–	53)
255	NCO$_2$C$_2$H$_5$	O	OAc	H	1)	5	7.5	2.5	–	–	55)

369	NCH$_3$	O	OAc	H	1)	5	5.5	3.5	—	—	55)
257†	O	S	OAc	H	14	4	8	1	—	—	27)
229†	O	NH	H	Br	14	—	—	—	7	8.5	58)
203†	O	NCH$_3$	OAc	H	13	5 1)	8	1	—	—	58)
300†			OH	H	13	—	8	1	—	—	58)
230†			H	Br	12	—	—	—	6	6	58)
266†	S	NCH$_3$	OAc	H	14	5.5	8	1	—	—	58)
301†			OH	H	14	6	8	1	—	—	58)
234†			H	Br	1)	—	—	—	7	7	58)
235†			H	Cl	12	—	—	—	7	7	58)
236†			H	OH	12	—	—	—	6	7	58)
370†			H	OAc	12	—	—	—	6	7.5	58)
371†			H	OTs	1)	—	—	—	7	7	58)
238†	OS$^{C(4)}$	NCH$_3$	H	Br	1)	—	—	—	8	8	58)
372†	OS$^{C(1)}$	NCH$_3$	H	Br	13	—	—	—	7	7	58)

1) Exact values could not be determined.

95

mined with the aid of spin,spin-decoupling experiments of the H–C(10)-signal. In Table 26 for such assignments characteristic coupling constants of a big variety of $C(10)^{X(2)}$- and $C(10)^{Y(7)}$-substituted 2,7-dihetero-twistanes as well as the calculated values for the unsubstituted 2,7-dioxa-twistane (*212*: see 5.2.3.2.1.2.) are listed.

5.2.3.2.2.4. 2-Oxa-7-thia-isotwistanes and -twistanes. Chemical shift values for the bridgehead H-atoms H–C(1), H–C(3), H–C(6) and H–C(8) as well as for H–C(10) at the substituent baring C-atom are listed in Table 27 for some isotwistanes and twistanes with X(2) = O, Y(7) = S, which belong to the three general types *G 92* $[C(10)^{O(2)}$-substituted 2-oxa-7-thia-isotwistanes], *G 93* $[C(10)^{S(7)}$-substituted 2-oxa-7-thia-isotwistanes] and *G 95* $[C(10)^{S(7)}$-substituted[t)] 2-oxa-7-thia-twistanes][uu)]. The data are subdivided into groups according to the nature of the substituents R at C(10). Within a certain group (equal substituent R) there are only small non-significant differences in chemical shifts between the signals for H–C(1) or the ones for H–C(6), however, each of the three H-atoms H–C(3), H–C(8) and H–C(10) exhibit characteristic positions of their signals. In compounds of the type *G 95* the signal for

	$R^{X(2)}$	$R^{Y(7)}$			$R^{X(2)}$	$R^{Y(7)}$
G 92	not H	H		*G 94*	not H	H
G 93	H	not H		*G 95*	H	not H

H–C(3) and H–C(8), resp., appear at higher field than in such of the types *G 92* and *G 93*. The signals for H–C(10) in compounds of the type *G 93* are at highest, in such of the type *G 95* at lowest field and in those of the type *G 92* between the two extremes. By comparing spectra of isomeric compounds with the same substituent R at C(10) belonging to the three types *G 92, G 93* and *G 95*, these differences in chemical shifts, together with some characteristic coupling constants (see Table 28) offer a suitable criterion for structural assignments.

Special attention should be paid to the coupling constants $J_{8,9}{}^{O(2)}$ and $J_9{}^{O(2)}$ $J_9{}^{O(2)}{}_{,10}{}^{O(2)}$. The values of 5.5 Hz for the former in isotwistanes *G 92* and *G 93* and of 1.5 Hz in twistanes *G 95* are in good agreement with the structures of such compounds. Especially in *G 95* the angle $\Phi_{8,9}{}^{O(2)}$ of approx. 65° as well as the trans-antiplanar arrangement of S(7) to H–C(9)$^{O(2)nn)}$ contribute to this small value. A coupling constant $J_9{}^{O(2)}{}_{,10}{}^{O(2)}$ of approx. 6.5–8.5 Hz, as observed in compounds of the type *G 95*, is characteristic for $C(10)^{Y(7)}$-substituted twistanes (see also the data in Table 26 on pp. 94 and 95.

5.2.3.2.3. C(4),C(10)-Disubstituted and C(4)- or C(10)-Monosubstituted 2,7-Dihetero-isotwistanes and -twistanes. The greatest number of comparable data of the many known C(4),C(10)-disubstituted ($R^1 + R^4$ or $R^2 + R^4$) and C(4)- or C(10)-

Table 27[53])

Type	R^X(2)	R^Y(7)	Compound	δH–C(1)	δH–C(3)	δH–C(6)	δH–C(8)	δH–C(10)O(2)	δH–C(10)S(7)
G 92	I	H	146	4.6–4.8	4.6–4.8	3.28	3.52	–	5.06
G 93	I	I	147	4.4–4.55	4.72	3.15–3.45	3.15–3.45	4.4–4.55	–
G 95	H	I	1)	1)	1)	1)	1)	1)	–
G 92	OAc	H	312	4.33	4.63	3.15–3.45	3.15–3.45	–	5.08
G 93	H	OAc	248	4.38	4.61	3.30	3.30	4.85	–
G 95	H	OAc	249	4.28	4.04	2.8–3.15	2.8–3.15	5.21	–
G 92	OH	H	317	4.26	4.58	3.11	3.28	–	3.95
G 93	H	OH	335	4.36	4.58	2.95	3.22	3.86	–
G 95	H	OH	364	4.0–4.4	3.91	2.9–3.2	2.9–3.2	4.0–4.4	–
G 92	OTs	H	314 2)	2)	2)	2)	2)	–	2)
G 93	·H	OTs	269	4.25	4.58	3.01	3.27	4.49	–
G 95	H	OTs	270	4.16	3.94	2.85–3.1	2.85–3.1	4.93	–

1) This compound has not been prepared.
2) 314 has not been isolated in pure state.

97

Table 28

Coupling constants [Hz]	$G\ 92^{1)}$ 146 $R^{X(2)} = I$	$G\ 93^{1)}$ 269 $R^{Y(7)} = OTs$	$G\ 95^{1)}$ 270 $R^{Y(7)} = OTs$
$J_{1,6}$			6
$J_{1,9O(2)}$		5.5	
$J_{1,10O(2)}$		5	6
$J_{3,8}$	6.5	6.5	
$J_{6,8}$	ca. 1		
$J_{6,10O(2)}$		2	
$J_{6,10S(7)}$	4		
$J_{8,9O(2)}$	5.5	5.5	1.5
$J_{9O(2),10O(2)}$			7.5

[1]) As the values are almost independent of the nature of the substituent R at C(10) for different derivatives of the same general type only one characteristic representative of each type (*G 92, G 93, G 95*) is listed.

G96   G97

monosubstituted (R^1 or R^2 or R^4) 2,7-dihetero-isotwistanes (*G 96*) and -twistanes (*G 97*) are available from acetoxy derivatives. Chemical shifts for the H-atoms geminal to the acetoxy group at C(4) and/or C(10), which are useful for the assignment of the configuration at these centers, are listed in Table 29 (isotwistanes) and Table 30 (twistanes). For a given combination of the heteroatoms X(2) and Y(7) in isotwistanes the signal for $H–C(10)^{X(2)}$ [$AcO–C(10)^{Y(7)}$] is at highest field, the one for $H–C(4)^{X(2)}$ [$AcO–C(4)^{Y(7)}$] at lowest field and the one for $H–C(10)^{Y(7)}$ [$AcO–C(10)^{X(2)}$] between the other two extremes. Data for $H–C(4)^{Y(7)}$ [$AcO–C(4)^{X(2)}$] are not yet available.

From the only directly comparable example in the twistane-series (*266* and *370*) most probably analogous trends can be expected. However, due to the higher symmetry, in twistanes only two different types of signals can be observed.

A special hint for the $C(4)^{Y(7)}$-configuration of the acetoxy group in isotwistanes and the $C(4)^{Y(7)}$- or $C(10)^{Y(7)t)}$-configuration in twistanes give the vicinal coupling constants of the H-atoms geminal to the acetoxy group, i.e. $H–C(4)^{X(2)}$ in isotwistanes and $H–C(4)^{X(2)}$ or $H–C(10)^{X(2)\ t)}$ in twistanes, with the three neighboring H-atoms: $J_4{}^{X(2)}{}_{,5}{}^{X(2)}$ [or $J_9{}^{X(2)}, 10^{X(2)}$] = approx. 5–8.5 Hz (dihedral angle approx. 20°), $J_{3,4}{}^{X(2)}$ [or $J_{1,10}{}^{X(2)}$] = approx. 4–6 Hz (dihedral angle approx. 40°) and $J_4{}^{X(2)}{}_{,5}{}^{Y(7)}$ [or $J_9{}^{Y(7)}{}_{,10}{}^{X(2)}$] = approx. 1–3.5 Hz [dihedral angle approx. 100° (see also Table 26 on pp. 94 and 95)].

Table 29. Isotwistanes (G 96)

Compound	Y(7)	X(2)	R^1–C(10)$^{X(2)}$	R^2–C(10)$^{Y(7)}$	R^3–C(4)$^{X(2)}$	R^4–C(4)$^{Y(7)}$	Refs.
				Substituents and δ [ppm]			
23	O	O	H: <4.7	OAc	H: 5.37	OAc	25)
246	O	O	H: 4.63	OAc	H	H	50)
304	O	O	I	H	H: 5.35	OAc	25)
305	O	O	OAc	H: 4.88	H: 5.33	OAc	25)
310	O	O	OAc	H: 4.86	H	H	50)
248	S	O	H: 4.85	OAc	H	H	53)
312	S	O	OAc	H: 5.08	H	H	53)
338	OS$^{C(1)bb}$	O	H: 4.98	OAc	H	H	54)
339	OS$^{C(1)bb}$	O	OAc	H: 5.32	H	H	54)
250	OS$^{C(4)bb}$	O	H: 5.17	OAc	H	H	54)
252	O$_2$S	O	H: 5.09	OAc	H	H	54)
78	NSO$_2$C$_6$H$_5$	O	H: 4.60	OAc	H: 5.28	OAc	36)
307	NSO$_2$C$_6$H$_5$	O	I	H	H: 5.20	OAc	35)
254	NCO$_2$C$_2$H$_5$	O	H: <4.8	OAc	H	H	55)
320	NCO$_2$C$_2$H$_5$	O	OAc	H: 4.77	H	H	55)
256	O	S	H: 4.80	OAc	H	H	27)
323	O	S	OAc	H: 5.14	H	H	27)
262	S	S	H: 5.10	OAc	H	H	57)
326	S	S	OAc	H: 5.44	H	H	27)
260	NCH$_3$	S	H: 4.79	OAc	H	H	58)
202	O	NCH$_3$	H: 4.45	OAc	H	H	58)
206	S	NCH$_3$	H: 4.84	OAc	H	H	58)
348	S	NCO$_2$C$_2$H$_5$	H: 4.70	OAc	H	H	58)

Table 30. Twistanes (*G 97*)

Compound	Y(7)	X(2)	Substituents and δ [ppm][t]		Refs.
			$R^1-C(10)^{X(2)}$ and/or $R^3-C(4)^{X(2)}$	$R^2-C(10)^{Y(7)}$ and/or $R^4-C(4)^{Y(7)}$	
243	O	O	H/H: 5.27/5.27	OAc/OAc	25)
247	O	O	H: 5.18	OAc	50)
249	S	O	H: 5.21	OAc	53)
251	OSC(4)bb)	O	H: 5.02	OAc	54)
253	O$_2$S	O	H: 5.04	OHc	54)
245	NSO$_2$C$_6$H$_5$	O	H/H 5.16/5.16	OAc/OAc	35)
255	NSO$_2$C$_6$H$_5$	O	H: 5.13	OAc	55)
369	NCH$_3$	O	H: 5.09	OAc	55)
257[t]	O	S	H: 5.22	OAc	27)
203[t]	O	NCH$_3$	H: 4.91	OAc	58)
266[t]	S	NCH$_3$	H: 5.09	OAc	58)
370[t]	S	NCH$_3$	OAc	H: 4.93	58)

5.2.3.3. 2,8-Dioxa-homotwistbrendanes[50)]

2,8-Dihetero-homotwistbrendanes (*G 4*) can be prepared from isotwistanes with a suitable leaving group $R-C(10)^{X(2)}$ by rearrangement involving neighboring group participation of Y(7) (see 3.4.3.). However, homotwistbrendanes with X(2) = O, Y(8) = O are the only compounds of this structural type known so far.

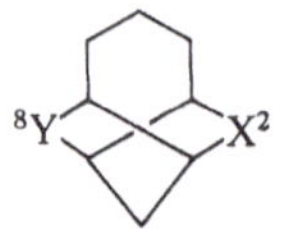

G4 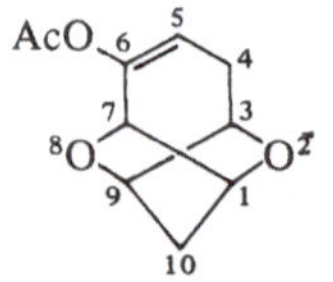　　　　　　　*378*

Table 31. Chemical shifts (δ [ppm])

H$_3$CCOO−C(6)	2.13
H−C(10) ⎱ AB-spectrum	1.99
H−C(10) ⎰	2.14
H−C(4) ⎱ AB-spectrum	2.5
H−C(4) ⎰	2.65
H−C(1) and H−C(3) (w$^{1/2}$ approx. 8)	4.1−4.3
H−C(9) (w$^{1/2}$ approx. 5)	4.50
H−C(7) (w$^{1/2}$ approx. 5)	4.73
H−C(5) (w$^{1/2}$ approx. 12)	5.37

Table 32. Coupling constants [Hz]

$J_{3,9}$	1.5	$J_{1,7}$	1.0	$J_{3,4}$	3
$J_{1,9}$	< 0.5	$J_{4,5}$	3.5	$J_{3,4'}$	1.5
$J_{9,10}$	1.5	$J_{4',5}$	5.0	$J_{1,10}$	0.75
$J_{9,10'}$	1.5	$J_{3,5}$	1.0	$J_{1,10'}$	0.5
$J_{5,7}$	< 0.5	$J_{4,4'}$	15.0	$J_{10,10'}$	10.0

The unsubstituted compound *377* [*G 4*: X(2) = O, Y(8) = O] (4.5.1.) exhibits in its NMR-spectrum two multiplets [δ = 4.10 ppm, $w^{1}/_2$ approx. 8 Hz and δ = 4.32 ppm, $w^{1}/_2$ approx. 5 Hz] in the ratio of 1 : 1, which belong to the four bridge-head H-atoms at C(1), C(3), C(7) and C(9). A clear proof for the 2,8-dioxa-homo-twistbrendane structure was obtained from the data of the enol-acetate *378* (4.5.1.). Chemical shifts are listed in Table 31 and coupling constants in Table 32. Each signal unequivocally could be assigned.

Acknowledgments. Our investigations were made possible by the generous support of the Schweizerischen Nationalfonds zur Förderung der wissenschaftlichen Forschung and by Ciba-Geigy AG, Basel, which is gratefully acknowledged.

6. References

[1] Vogt, B. R.: Tetrahedron Letters *1968*, 1575.
[2] Whitlock, Jr., H. W., Siefken, M. W.: J. Amer. Chem. Soc. *90*, 4929 (1968).
[3] Baldwin, J. E., Foglesong, W. D.: J. Amer. Chem. Soc. *90*, 4303 (1968).
[4] Ganter, C., Moser, J.-F.: Helv. *52*, 725 (1969).
[5] Lunn, W. H. W.: J. Chem. Soc. (C) *1970*, 2124.
[6] Spurlock, L. A., Clark, K. P.: J. Amer. Chem. Soc. *94*, 5349 (1972).
[7] Schleyer, P. von R., Wiskott, E.: Tetrahedron Letters *1967*, 2845.
[8] Dickinson, R. G., Raymond, A. L.: J. Amer. Chem. Soc. *45*, 22 (1923).
[9] Gonell, H. W., Mark, H.: Z. physikal. Chem. *107*, 181 (1923).
[10] Stetter, H.: Angew. Chem. *66*, 217 (1954).
[11] Stetter, H.: Angew. Chem. *74*, 361 (1962).
[12] Fort, Jr., R. C., Schleyer, P. von R.: Chem. Reviews *64*, 277 (1964).
[13] Bingham, R. C., Schleyer, P. von R.: Fortschr. Chem. Forschung *18*, 1 (1971).
[14] McKervey, M. A.: Chem. Soc. Reviews *3*, 479 (1974).
[15] Landa, S., Macháček, V.: Coll. Czechoslov. Chem. Commun. *5*, 1 (1933).
[16] Landa, S.: Chem. Listy *27*, 415 (1933).
[17] Prelog, V., Seiwerth, R.: Ber. deutsch. chem. Ges. *74*, 1644, 1769 (1941).
[18] Schleyer, P. von R.: J. Amer. chem. Soc. *79*, 3292 (1957).
[19] Gelbard, G.: Ann. Chim. (Paris) *1969*, 331.
[20] Stetter, H., Meissner, H.-J.: Tetrahedron Letters *1966*, 4599.
[21] Stetter, H., Meissner, H.-J., Last, W.-D.: Chem. Ber. *101*, 2889 (1968).
[22] Cuthbertson, E., MacNicol, D. D.: Tetrahedron Letters *1974*, 2367.
[23] Averina, N. V., Zefirov, N. S., Kadzyauskas, P. P., Rogozina, S. V., Sadovaya, N. K., Solda-tov, N. M.: Zh. Org. Khim. *10*, 1442 (1974).
[24] Wicker, N. I.: Diss. ETHZ, Nr. 4672 (1972).
[25] Ackermann, P.: Diss. ETHZ, Nr. 5354 (1974).
[26] Wirthlin, T., Ganter, C.: unpublished results.
[27] Capraro, H.-G., Ganter, C.: unpublished results.
[28] Ganter, C., Wicker, K.: Helv. *51*, 1599 (1968).
[29] Wigger, N., Ganter, C.: unpublished results.
[30] Stetter, H., Mehren, R.: Liebigs Ann. Chem. *709*, 170 (1967).
[31] Stetter, H., Heckel, K.: Tetrahedron Letters *1972*, 801.
[32] Stetter, H., Heckel, K.: Chem. Ber. *106*, 339 (1973).
[33] Kashman, Y., Benary, E.: J. org. Chemistry *37*, 3778 (1972).
[34] Ganter, C., Portmann, R. E.: Chimia *25*, 246 (1971).
[35] Portmann, R. E.: Diss. ETHZ, Nr. 4933 (1974).
[36] Portmann, R. E., Ganter, C.: Helv. *56*, 1962 (1973).
[37] Kashman, Y., Benary, E.: Tetrahedron *28*, 4091 (1972).
[38] Blanc, P. Y., Diehl, P., Fritz, H., Schläpfer, P.: Experientia *23*, 896 (1967).
[39] Lautenschlaeger, F.: J. org. Chemistry *34*, 4002 (1969).
[40] Ganter, C., Portmann, R. E.: Helv. *54*, 2069 (1971).
[41] Rassat, A.: Pure Appl. Chemistry *25*, 623 (1971).
[42] Dupeyre, R.-M., Rassat, A.: Tetrahedron Letters *1973*, 2699.
[43] Stetter, H., Heckel, K.: Tetrahedron Letters *1972*, 1907.
[44] Portmann, R. E., Ganter, C.: Helv. *56*, 1986 (1973).
[45] Ganter, C., Wicker, K.: Helv. *53*, 1693 (1970).
[46] Ganter, C., Wicker, K., Wigger, N.: Chimia *24*, 27 (1970).
[47] Wigger, N., Wicker, K., Zwahlen, W., Ganter, C.: Chimia *25*, 418 (1971).
[48] Duthaler, R. O., Wicker, K., Ackermann, P., Ganter, C.: Helv. *55*, 1809 (1972).
[49] Ackermann, P., Tobler, H., Ganter, C., Helv. *55*, 2731 (1972).
[50] Wicker, K., Ackermann, P., Ganter, C.: Helv. *55*, 2744 (1972).
[51] Ackermann, P., Tobler, H., Ganter, C.: Chimia *26*, 658 (1972).
[52] Duthaler, R. O.: Diss. ETHZ, Nr. 5108 (1973).

[53] Wigger, N., Ganter, C.: Helv. *55*, 2769 (1972).
[54] Wigger, N., Stücheli, N., Szczepanski, H., Ganter, C.: Helv. *55*, 2791 (1972).
[55] Portmann, R. E., Ganter, C.: Helv. *56*, 1991 (1973).
[56] Ganter, C., Moser, J.-F.: Helv. *52*, 725 (1969).
[57] Ganter, C., Wigger, N.: Helv. *55*, 481 (1972).
[58] Szczepanski, H.: Diss. ETHZ, Nr. 5576 (1975).
[59] Whitlock, Jr., H. W.: J. Amer. Chem. Soc. *84*, 3412 (1962).
[60] Ganter, C., Zwahlen, W.: Helv. *54*, 2628 (1971).
[61] Ackermann, P., Ganter, C.: Helv. *56*, 3054 (1973).
[62] Cope, A. C., McKervey, M. A., Weinshenker, N. M.: J. org. Chemistry *34*, 2229 (1969).
[63] Cuthbertson, E., MacNicol, D. D.: J. chem. Soc., Perkin I, *1974*, 1893.
[64] Ganter, C., Wicker, K., Zwahlen, W., Schaffner-Sabba, K.: Helv. *53*, 1618 (1970).
[65] Dittmann, W., Sunder-Plassmann, P.: Chemiker Ztg. *94*, 299 (1970).
[66] Inhoffen, H. H., Kölling, G., Koch, G., Nebel, I.: Chem. Ber. *84*, 361 (1951).
[67] Capon, B.: Quart. Reviews *18*, 45 (1964).
[68] Lwowski, W.: Angew. Chem. *70*, 483 (1958).
[69] Streitwieser, Jr., A.: Chem. Reviews *56*, 571 (1956).
[70] Zwahlen, W.: Diss. ETHZ, Nr. 4949 (1973).
[71] Ganter, C., Duthaler, R. O., Zwahlen, W.: Helv. *54*, 578 (1971).
[72] Perst, H.: Oxonium ions in organic chemistry. Academic Press: Verlag Chemie 1971.
[73] Gundermann, K.-D.: Angew. Chem. *75*, 1194 (1963).
[74] Mueller, W. H.: Angew. Chem. *81*, 475 (1969).
[75] Helmkamp, G. H., Owsley, D. C.: Mechanisms of Reactions of Sulfur Compounds *4*, 37 (1969).
[76] Kwart, H., Drayer, D.: J. org. Chemistry *39*, 2157 (1974).
[77] Ikegami, S., Ohishi, J., Shimizu, Y.: Tetrahedron Letters *1975*, 3923.
[78] Bordwell, F. G., Brannen, Jr., W. T.: J. Amer. Chem. Soc. *86*, 4645 (1964).
[79] Hammer, C. F., Heller, S. R., Craig, J. H.: Tetrahedron *28*, 239 (1972).
[80] Tichý, M., Sicher, J.: Tetrahedron Letters *1969*, 4609.
[81] Capraro, H.-G., Ganter, C.: Helv. *59*, 97 (1976).
[82] Gerlach, H.: Helv. *51*, 1587 (1968).
[83] Dale, I. A., Dull, D. L., Mosher, H. S.: J. org. Chemistry *34*, 2543 (1969).
[84] Tichý, M.: Tetrahedron Letters *1972*, 2001.
[85] Adachi, K., Naemura, K., Nakazaki, M.: Tetrahedron Letters *1968*, 5467.
[86] Cope, A. C., Anderson, B. C.: J. Amer. Chem. Soc. *79*, 3892 (1957).
[87] Kimura, K., Nagakura, S.: Spectrochim. Acta *17*, 166 (1961).
[88] Ganter, C., Moser, J.-F.: Helv. *54*, 2228 (1971).
[89] Procházka, M., Palaček, M.: Coll. Czechoslov. Chem. Commun. *32*, 3049 (1967).
[90] Fehnel, C. A., Carmack, M.: J. Amer. Chem. Soc. *71*, 84 (1949).
[91] Bergson, G., Delin, A.-L.: Arkiv Kemi *18*, 489 (1961).
[92] Bergson, G., Claeson, G., Schotte, L.: Acta chem. scand. *16*, 1159 (1962).
[93] Ganter, C., Moser, J.-F.: Helv. *51*, 300 (1968).
[94] Williams, D. H., Fleming, I.: "Spektroskopische Methoden in der Org. Chemie", Georg Thieme-Verlag (1968).
[95] Kessler, H.: Angew. Chem. *82*, 237 (1970).
[96] La Planche, L. A., Rogers, M. T.: J. Amer. Chem. Soc. *85*, 3728 (1963).
[97] Castellano, S., Bothner-By, A. A.: J. chem. Phys. *41*, 3863 (1964).
[98] Foster, A. B., Inch, I. D., Qadir, M. G., Webber, J. M.: Chem. Commun. *1968*, 1086.
[99] Cooper, R. D. G., De Marco, P. V., Cheng, J. C., Jones, N. D.: J. Amer. Chem. Soc. *91*, 1408 (1969).
[100] Hamon, A., Lacoume, B., Olivie, J.: Bull. Soc. Chim. France *1971*, 1472.
[101] Sternhall, S.: Rev. Pure Appl. Chemistry *14*, 15 (1964).

7. Notes

a)   The capital letter G in front of a number shall characterize a general formula, to distinguish them from the set of specific compounds.

b)   Adamantane: tricyclo[3.3.1.1^{3,7}]decane.

c)   For tricyclo[4.3.1.0^{3,8}]decane the following trivial names were introduced in the literature: isoadamantane (1968)[1], protoadamantane (1968)[2,3], isotwistane (1969)[4] and 2(3 → 4)abeo-adamantane (1970)[5]. In the present review the name isotwistane will be applied, which until now is exclusively used to describe such 2,7-dihetero-tricyclodecanes. However, one should pay attention to the fact that since 1972[6] the trivial name isotwistane is also applied for tricyclo[4.3.1.0^{3,7}]decanes.

d)   Twistane[2]: tricyclo[4.4.0.0^{3,8}]decane.

e)   In accordance to the known twistbrendane (G 6)[7], tricyclo[5.3.0.0^{3,9}]decanes will be called homotwistbrendanes.

$G6$

f)   Tricyclo[3.3.2.0^{3,7}]decane: octahydro-2,5-ethanopentalene.

g)   See *e.g.* the review articles by Stetter[10,11], Fort, Jr. and Schleyer[12], Bingham and Schleyer[13] and McKervey[14] as well as references cited therein.

h)   See the review articles by Stetter[10, 11] and Gelbard[19].

i)   See 3.2.2.2. for a detailed discussion.

j)   In isotwistanes and twistanes the indices X(2) and Y(7) indicate the bridge-heteroatoms toward which a substituent is orientated.

k)   For an independant synthesis of 9 and 23 from diiodide 5, see 3.2.2.

l)   In adamantanes the indices X(2) and Y(6) indicate the bridge-heteroatoms towards which a substituent is orientated.

m)   See 3.2.2. for a detailed discussion.

n)   See the detailed discussion in 3.2.2. (for 47), in 3.4.2 (for 50) and in 3.3.1. (for 54).

o)   For another preparation of 48 and 52 see 2.1.5.3., of 49 see 2.1.5.4.

p)   For an independant synthesis of 49 and 78 from diiodide 47, see 3.2.2.2.

q)   In this connection it should be mentioned that bromodemercuration of 121 under analogous conditions predominantly yielded the 10$^{O(7)}$-bromide 126.

r)   Analogous experiments starting from 139 with mercuric acetate in water or diluted acetic acid (→ 141) followed by NaBH$_4$-reduction in basic solution, led also to 148, although in much lower yield (< 20%).

s)   In 1970 a paper appeared by Dittmann and Sunder-Plassmann[65]. These authors are supposed to have obtained 2,7-dioxa-twistane (212) by an independent route. However, from the experiments they described it seems very unlikely to be the case.

t)   To facilitate comparisons between isotwistanes, twistanes and homotwistbrendanes the atoms in most substituted twistanes and homotwistbrendanes are numbered against IUPAC-rules in such a way, that in twistanes a substituent correctly located at C(4) or C(5) will be placed at C(10) and in homotwistbrendanes a substituent correctly located at C(4) will be placed on C(6). By consequence in several cases the two heteroatoms X and Y became numbered against IUPAC-rules, too.

u)   For reviews on this subject, see[67−69].

v)   Extensive studies are also known on substitutions and rearrangements involving neighboring group participation of the heteroatom Y(9) in 9-heterobicyclo[3.3.1]- and 9-heterobicyclo[4.2.1]nonanes (see the onium ion G 35), *e.g.*[24, 60, 64, 70, 71].

G 35

w) On oxonium ions, see[72].
x) On episulfonium ions, see[73–77].
y) In contrast to sulfides, sulfoxides and sulfones show no tendency of neighboring group participation in exchange reactions at the β-carbon atom, see[78] as well as experiments on 9-thiabicyclo[3.3.1]nonenes: treatment of *239*, *240* and *241* with silver acetate in acetic acid[54].

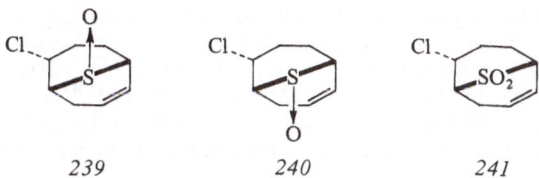

| *239* | *240* | *241* |

z) On aziridinium ions, see [79] and references cited therein.
aa) In none of the studied cases products from an attack C, F, I and L could be observed. Therefore these pathways have been omitted in the above general scheme.
bb) In 7-thia-isotwistane 7-oxides the indices C(1) and C(4) indicate the C-atoms towards which the oxygen at S(7) is orientated.
cc) The opposite arrangement, R^1 anti to Y and R^2 anti to X(2), leads to the same disubstituted products *G 73 – G 80*, because in *G 73* and *G 77*, in *G 74* and *G 79*, in *G 75* and *G 78* as well as in *G 76* and *G 80* in each case the two substituents R^3 and R^4 are only interchanged.
dd) In none of the studied cases products from an attack of an external nucleophile $R^{3\ominus}$ or $R^{4\ominus}$ like C, F, I, L, O, R, U and X could be observed. Therefore these pathways were omitted in the above general scheme.
ee) An analogous result was obtained with the corresponding $10^{S(2)}$-tosyloxy-2-thia-7-oxa-isotwistane (*343*), see 4.3.4.
ff) Numbering according to the isotwistane skeleton.
gg) The indices C(5) and C(9) indicate the carbon atoms towards which the oxygen at S(2) is orientated.
hh) To differenciate the optically active compounds from the racemic ones described in the other sections, the former will be numbered differently. The numbers of the corresponding racemic compounds are added in brackets in the formula schemes.
ii) CD-measurements on the (−)-diene *401* by Prof. Dr. G. Snatzke (Ruhr-Universität Bochum, Germany), which are thankfully acknowledged, are in full agreement with the above assigned absolute configuration.
jj) Compare isopropyl iodide: λ_{max} = 259.4 nm (ϵ = 566) in CH_3OH[87] and 2-iodo-9-oxa-bicyclo[3.3.1]nonane-derivatives: λ_{max} = approx. 260 nm (ϵ approx. 630) in C_2H_5OH (see 5.2.2.1.).
kk) All the chemical shifts (δ [ppm]) discussed in 5.2.3.1. corresponding to AB-systems of the methylene groups $H_2-C(4)$, $H_2-C(8)$, $H_2-C(9)$ and $H_2-C(10)$ are corrected values according to the formula $\delta_A - \delta_B = \sqrt{(\gamma_4 - \gamma_1) \cdot (\gamma_3 - \gamma_2)}$ (see [94]).
ll) For some special discussions of nmr.-data in this section in some compounds the substituents at C(4) and C(8) are interchanged compared to the numbering in preceeding sections.
mm) As the substituents at C(4) and C(8) do not have an essential influence on the mentioned vicinal coupling constants, they have been omitted in the formula *G 86 – G 89*.
nn) See the detailed discussion of these effects in [94]

oo) As the exact positions of the signals for $H-C(9)^{O(2)}$ and $H-C(9)^{N(6)}$ could not be obtained, the studies are restricted to $H-C(10)^{O(2)}$ and $H-C(10)^{N(6)}$.

pp) Corresponding results are observed for $H_2-C(9)$ in $4^{O(2)},8^{N(6)}$-disubstituted compounds.

qq) *E.g.* A = $H-C(4)^{O(2)}$ or $H-C(9)^{O(2)}$, A' = $H-C(5)^{O(7)}$ or $H-C(10)^{O(7)}$, B = $H-C(4)^{O(7)}$ or $H-C(9)^{O(7)}$, B' = $H-C(5)^{O(2)}$ or $H-C(10)^{O(2)}$. $H-C(4)^{O(2)}$ and $H-C(9)^{O(7)}$ etc. each are equivalent because of the D_2-symmetry in *212*.

rr) Compare also the coupling constants of the C(10)-substituted 2,7-dioxa-twistanes *247*, *268*, *355* and *359* in table 26 (5.2.3.2.2.3.).

ss) On the basis of model studies small values for the coupling constants $J_{1,9}Y(7)$ and $J_{8,9}Y(7)$ can be expected from the estimated values for the corresponding angles. Furthermore the effect of the electronegative substituents X(2) and Y(7) to diminish J_{vic} has to be taken into account, also (see [94]) and the references cited therein).

tt) $J_5O(2),6$ is almost zero. This is in good agreement with the angle between the corresponding H-atoms of approx. 90° and the influence of O(7) on J_{vic} (see [94]) as well as [ss])).

uu) Comparisons with compounds of the fourth type *G 94* ($R^{X(2)} = H$, $R^{Y(7)} = H$) are less available because in such compounds (*e.g.* X(2) = NCH$_3$, Y(7) = $S^{t)}$: *235*[58], *236*[58], *370*[58] etc.) the signals for the four bridgehead H-atoms are usually one broad unresolved multiplet, wherefore the chemical shifts have not been determined for each single H-atom.

Received April 2, 1976

Olefin Insertion in Transition Metal Catalysis

Dr. Gisela Henrici-Olivé and Prof. Dr. Salvador Olivé

Monsanto Research S. A., CH-8050 Zürich, Switzerland*)

Contents

*) Present address: Monsanto Triangle Park Development Center Inc., P.O. Box 12274, Research Triangle Park, N.C. 27709, U.S.A.

G. Henrici-Olivé and S. Olivé

1. Introduction

Homogeneous catalysis with defined soluble transition metal complexes as catalysts has become one of the most effective means of transforming simple olefins into more valuable materials. The technically important hydroformylation of olefins to aldehydes or alcohols[1], the Wacker process[2], the dimerization of propylene to linear hexenes[3], the oligomerization of ethylene to linear α-olefins[4], are only a few examples. A feature common to all these processes is the insertion of a substrate olefin molecule, which is coordinatively bonded to the transition metal center M, into a metal-carbon or metal-hydrogen bond present at the same center:

$$\overset{\diagup}{\underset{\diagdown}{C}}=\overset{\diagup}{\underset{\diagdown}{C}} \quad\quad | \quad | $$

$$M\!-\!R \longrightarrow M\!-\!\overset{|}{\underset{|}{C}}\!-\!\overset{|}{\underset{|}{C}}\!-\!R \tag{1}$$

(R = alkyl group or H)

If the olefin is unsymmetrically substituted, the structure of the product will evidently depend on the direction of insertion; thus, for propylene the two insertion modes (2) and (3) may in principle be formulated:

$$H_2C=CH\!-\!CH_3$$

$$M\!-\!R \longrightarrow M\!-\!CH_2\!-\!\underset{\underset{CH_3}{|}}{CH}\!-\!R \tag{2}$$

$$CH_3\!-\!CH=CH_2$$

$$M\!-\!R \longrightarrow M\!-\!\underset{\underset{CH_3}{|}}{CH}\!-\!CH_2\!-\!R \tag{3}$$

Markownikoff's rule is frequently applied to this problem, and one of the insertion modes is designated as Markownikoff mode, and the other as "anti-Markownikoff" mode. There is, however some confusion in the literature as to which is which.

Markownikoff's rule states that during the addition of H^+X^- to an asymmetrically substituted olefin the negative part becomes attached to the unsaturated carbon carrying the smaller number of hydrogen atoms[a]. Originally this was an empirical rule, but in the meantime quantum chemistry and molecular orbital theory have provided a solid theoretical background to it. In the particular case of propylene, for instance, it has been calculated that the π-electron density distribution is as follows[6]:

$$\overset{0.972 \quad 1.042}{CH_3\!-\!CH=CH_2}$$

---

[a] The original "Markownikoff rule" (1875) reads: "Lorsqu'à un hydrocarbure non saturé, renfermant des atomes de carbone inégalement hydrogénés, s'ajoute un acide haloidhydrique, l'element électronegatif se fixe sur le carbone le moins hydrogené"[5].

i.e. there is a certain polarization of the double bond, and one expects an anion to attack at the carbon atom next to the methyl group.

We define then as the Markownikoff mode of insertion that expected from a consideration of the electronic structure of the reaction partners in their ground state. With this definition in mind we shall now first discuss briefly the polarity of metal-carbon and metal-hydrogen bonds, as well as the π-electron distribution in certain relevant olefins. We shall then proceed to a description of numerous Markownikoff and anti-Markownikoff insertion reactions cited in the literature, and we shall try to discern electronic and steric effects leading to the one or to the other.

2. Electronic Structure of Reaction Partners

2.1. Polarity of M–C and M–H Bonds

There seems to be general agreement concerning the character of metal-carbon bonds in transition metal complexes having alkyl groups linked to the metal: the α-carbon atom of the alkyl group and the metal form a normal covalent σ-bond, each partner contributing one electron. From X-ray structural analyses of such complexes it is known that the bond lengths are very nearly those predicted by addition of the covalent radii for carbon and the transition metal in the appropriate valence state[7]. Nevertheless the metal-carbon bonds are doubtlessly somewhat polarized, in the sense that the transition metal cation has a certain positive, the α-carbon of the alkyl group a negative charge. Several molecular orbital calculations of varying levels of sophistication have indicated that the positive charge on the metal center is mostly between $+1$ and $+2$, independently of the formal oxidation state of the metal[8, 9]. The negative charge on the α-carbon has also been computed, for the particular case of titanium based Ziegler type catalysts for the polymerization and oligomerization of ethylene[8, 10]. Charges of -0.24 to -0.32 have been reported.

The polarization of the metal-carbon bonds in the indicated direction may also be inferred from direct experimental evidence. Nearly all alkyl transition metal compounds are cleaved by protonic solvents or acids, and in all cases the proton ends up with the alkyl group giving an alkane, whereas the anionic part of the cleaving agent adds to the metal[11, 12]. A few examples are given below:

$$CH_3Ni\,(h^3-C_3H_5)L + HBr \longrightarrow CH_4 + BrNi\,(h^3-C_3H_5)L$$

$$[L = P(C_6H_5)_3;^{13)}]$$

$$(CH_3)_2PdL_2 + C_2H_5OH \longrightarrow CH_4 + CH_3(C_2H_5O)PdL_2$$

$$[L = P(C_2H_5)_3;^{11)}]$$

$$L_n V(C_3H_6)_p R + {}^3HX \longrightarrow {}^3H(C_3H_6)_p R + L_n VX$$

L_n = usual ligands of Ziegler type polymerization catalyst;
C_3H_6 = propylene; R = initiating group; ${}^3HX$ = tritiated Brønsted acid[14].

109

From the sum of these evidences we may safely conclude that there is a positive charge on the metal, and a moderate negative charge on the carbon.

The hydrogen in transition metal-hydrogen bonds is generally assumed to have a negative charge too, and the corresponding complexes are called hydride- or hydrido-transition metal complexes. This name, although commonly used, is not quite correct in view of the covalent character of the M—H bond. Older X-ray studies did not permit to locate exactly the hydrogen in the vicinity of a heavy metal atom, but with the advent of neutron diffraction and high precision data collection, metal-hydrogen distances could be determined with sufficient precision to permit the statement that the hydrogen is situated close to what might be calculated as the normal covalent bond distance[15]. ESR data may also be quoted as a corroboration of the covalent character of M—H bonds, although such data are available only for a few paramagnetic complexes such as $[Cp_2 TiH]_2^-$ [16], $[Cp_2 Ti(H)(H)]^-$ [17], and $HCoR(L-L)$ [18] $[Cp = h^5 —C_5H_5; R = methyl or ethyl; (L-L) = (C_6H_5)_2 PCH_2 CH_2 P(C_6H_5)_2$ or $(C_2H_5)_2 PCH_2 CH_2 P(C_2H_5)_2]$. In all these cases, the coupling of the unpaired electron of the transition metal ion with the hydrogen nuclei provides evidence for covalent bonding.

The most frequently cited experimental evidence for a negative charge on the hydrogen is the large high field shift in the nuclear magnetic resonance spectra of protons directly bonded to transition metals, in diamagnetic complexes (τ between 10 and 50 ppm). This shift is even used as a criterium of the presence of a hydrido complex, particularly in species obtainable only in solution. Several explanations for the high values of the shift have been put forward[19]. An early suggestion was that the shift should be entirely caused by a strong local shielding of the proton, produced by a rather ionic bond with hydride-like character. However, τ for the free H^- ion is only 5 ppm, so τ values up to 50 ppm cannot be entirely due to such local shielding. Ligand influenced electron population in the large and diffuse 4s and 4p orbitals of (first row) transition metals[20], as well as effects of a distortion of the partly filled d-shell by the magnetic field[19] have been suggested as the main contributors to the shift. But, nevertheless, the theoretical work points also to a negative charge at the hydrogen. Thus, LCAO—MO calculations of the charge distribution in a series of hydrido complexes of Cr, Mo, W, Fe, Co, Rh and Ir have been used to calculate the chemical shifts. Computed and observed shift data are compatible only when the H atom is assumed to have an excess negative charge, corresponding to a total of about 1.1 to 1.4 electrons around the proton[21]. Moreover, a recent MO calculation of a hydrido-platinum(II) complex places a charge of 0.178 electrons on the hydrogen[22].

Further chemical evidence for the polarization of the metal-hydrogen bonds as $M^{\delta +}$—$H^{\delta -}$ comes from cleavage reactions with aqueous acids, alcohols or phenols to produce H_2, or with halogen (Br_2 or I_2) to give hydrogen halide, although these reactions are sometimes not quantitative. Some examples are[23]:

$$HRhL_3 + C_6H_5OH \xrightarrow{\text{toluene}} Rh(OC_6H_5)L_3 + H_2$$

$$HRuCl(CO)L_3 + HNO_3 \xrightarrow{C_2H_5OH} RuCl(NO_3)(CO)L_3 + H_2$$

$$H_2IrClL_3 + HCl \xrightarrow{C_2H_5OH} HIrCl_2L_3 + H_2$$

$$HRhBr_2L_3 + Br_2 \xrightarrow{CH_2Cl_2} RhBr_3L_3 + HBr$$

(L = phosphine ligands).

In a compound which contains both an alkyl and a hydride ligand, the latter is even cleaved in preference by HCl[24]:

$$CH_3RuH(L-L) + HCl \longrightarrow CH_3RuCl(L-L) + H_2$$

A distinct resemblance in character of M—C and M—H σ-bonds is manifest in the similar *trans*-influence exerted by alkyl and hydride ligands, *e.g.* in the square-planar complexes *trans*-$CH_3PtCl(PR_3)_2$ and *trans*-$HPtCl(PR_3)_2$, as borne out by the Pt—Cl stretching frequency. Table 1 indicates, on the left hand side, the high *trans*-influence (*i.e.* the marked electron donor property) of CH_3 and H, as compared with other ligands. Moreover, the strengths of metal-carbon and metal-hydrogen bonds are influenced in a comparable way by the electronic properties of the ligand *trans* to them. As can be seen in the middle and on the right hand side of Table 1, the stretching frequencies of the Pt—C as well as those of the Pt—H bond decrease (*i.e.* the bonds are increasingly weakened) with increasing *trans*-influence of the anionic ligand.

Table 1. *Trans*-influence in several square-planar Pt(II) complexes (L = phosphine)

trans-$XPtClL_2$ [25]		*trans*-CH_3PtXL_2 [26]		*trans*-$HPtXL_2$ [27]	
X	$\nu_{Pt-Cl}(cm^{-1})$	X	$\nu_{Pt-C}(cm^{-1})$	X	$\nu_{Pt-H}(cm^{-1})$
H	269	CN	2041	CN	516
CH_3	274	I	2156	I	540
$P(C_2H_5)_3$	295	Br	2178	Br	548
$P(C_6H_5)_3$	298	Cl	2183	Cl	551
$P(OC_6H_5)_3$	316	NO_3	2242	NO_3	566
Cl	340				
CO	344				

There seems to be some confusion about the polarization of the metal-hydrogen bond in certain hydrido-transition metal-carbonyl compounds, in particular $HCo(CO)_4$. This and some related hydrido-carbonyl compounds are slightly soluble in water, and such solutions are clearly acidic[23]. In the gas phase or in non-polar solvents, however, these complexes exhibit spectroscopic and chemical properties similar to those of other "non-acidic" hydride complexes. A MO calculation of the charge densities in $HCo(CO)_4$ indicated that 1.6 electrons are associated with the hydrogen atom, and the question was discussed as to how a molecule with a negatively charged hydrogen can dissolve in water to become an acid[28]. It was estimated that the process

$$HCo(CO)_4 \ (g) \longrightarrow [Co(CO)_4]^- \ (aq) + H^+ \ (aq)$$

would release about 135 kcal/mol, mainly due to the high solvation energy of the proton and of the anionic metal species, and hence "has a strong motivation to take place". Evidently, an acidic dissociation of a hydrido-transition metal complex is feasible only if the negative charge on the anionic part can be dissipated over several strongly electron accepting ligands, in particular CO ligands. In such cases, and with powerfully solvating solvents such as water or alcohols, one may then have to take into consideration a reversal of the polarization of M–H bonds, and also of certain M–C bonds. Thus, H_2O and alcohols cleave acyl-metal bonds in certain transition metal carbonyl complexes to give carboxylic acid derivatives and a metal-hydride, e.g.:

$$RCCo(CO)_4 + C_2H_5OH \ \rightarrow RCOC_2H_5 + HCo(CO)_4$$
$$\overset{\|}{O} \qquad\qquad\qquad\qquad \overset{\|}{O}$$

This is actually a crucial step in the carboxylation of olefins (Reppe reaction) with transition metal carbonyl catalysts[29]. Apart from these extreme cases, however, metal-carbon σ-bonds as well as metal-hydride bonds are best described as moderately polarized covalent bonds, with negative charge on the carbon or hydrogen respectively:

$$\overset{\delta +}{M} - \overset{\delta -}{C}\!\!\lesseqgtr \qquad\qquad \overset{\delta +}{M} - \overset{\delta -}{H}$$

2.2. Polarity of Olefinic Double Bonds

If one compares the electronic situation of the double bond of a mono-substituted ethylene such as propylene or acrylonitrile with that of ethylene itself, one has to consider two influences of the substituent. On one side, electron density may be donated to, or withdrawn from, the unsaturated hydrocarbon fragment as a whole; on the other side the distribution of electrons remaining in the hydrocarbon fragment may be polarized so that electrons are drawn to or from the site of substitution. According to the usual nomenclature the first is the inductive, the second the resonance effect of the substituent. This separation has been made originally on empirical grounds by Taft et al.[30]. Later, a satisfying substantiation has resulted from the calculation of charge density distributions in the framework of MO theory (CNDO/2, or complete neglect of differential overlap approximation)[6]. The double classification of substituents according to the mentioned criteria may be expressed by +I and –I for the inductive effect, and by a – or + superscript for the polarization of the double bond, for higher electron density at the substituted or unsubstituted

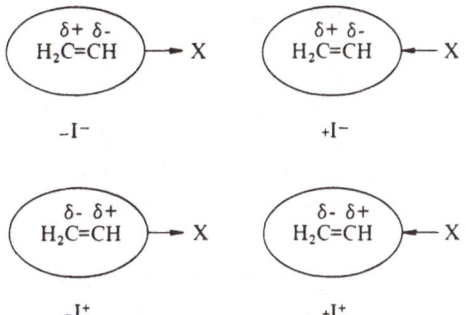

Fig. 1. Schematic representation of types of inductive substituents X in mono-substituted ethylene (after[6])

Table 2. Substituent effects in $H_2C=CH-X$[6, 30, 32)]

X	Classification	Polarization of the double bond
CH_3	$+I^+$	$H_2\overset{\delta-}{C} = \overset{\delta+}{CH} - CH_3$
C_6H_5	$-I^+$	$H_2\overset{\delta-}{C} = \overset{\delta+}{CH} - C_6H_5$
Cl	$-I^+$	$H_2\overset{\delta-}{C} = \overset{\delta+}{CH} - Cl$
F	$-I^+$	$H_2\overset{\delta-}{C} = \overset{\delta+}{CH} - F$
OCH_3	$-I^+$	$H_2\overset{\delta-}{C} = \overset{\delta+}{CH} - OCH_3$
CF_3	$-I^-$	$H_2\overset{\delta+}{C} = \overset{\delta-}{CH} - CF_3$
CN	$-I^-$	$H_2\overset{\delta+}{C} = \overset{\delta-}{CH} - CN$
COOH	$-I^-$	$H_2\overset{\delta+}{C} = \overset{\delta-}{CH} - COOH$
COOR	$-I^-$	$H_2\overset{\delta+}{C} = \overset{\delta-}{CH} - COOR$
$COCH_3$	$-I^-$	$H_2\overset{\delta+}{C} = \overset{\delta-}{CH} - COCH_3$

carbon atom respectively. This leads to the four types of substituents sketched in Fig. 1.

Among the substituents relevant in our context (see Table 2), only alkyl groups are of the +I type, although the amount of electron density actually donated to the double bond is relatively small, as may be seen from the example of propylene as given in the Introduction. All other substituents withdraw electron density from the $H_2C=CH$- fragment (—I). More important for our purpose, however, is the polarization of the double bond. It has been shown[6, 30)] that those substituents which have the most electronegative atom directly attached to the hydrocarbon fragment draw electron density from the site of substitution to the CH_2 group, whereas substituents having the electronegative atom one position further away polarize the double bond in the opposite direction. In Table 2, the substituents of each group are ordered approximately with increasing polarizing effect[30)].

In the case of styrene, simple Hückel MO theory would indicate equal π-electron density (=1) at the α and β carbon atom[31]; more sophisticated calculations, however, reveal a polarization as shown in Table 2[32], in agreement with Taft's indices[30].

In the MO treatment of chemical reactivity, the overlap of the highest occupied molecular orbital (HOMO) of one reaction partner with the lowest unoccupied molecular orbital (LUMO) of the other partner plays an important role. In a very rough, qualitative manner we may represent the HOMO's and LUMO's of ethylene and substituted ethylenes as shown in Fig. 2[b].

The energy levels of the HOMO's are known from photoelectron spectroscopy, those of the LUMO's are estimated from electron affinities, substituent effects on charge transfer spectra, polarographic reduction potentials, and electron ($\pi-\pi^*$) absorption spectra[34]. Typical values are summarized in Table 3. The energies of the HOMO's of substituted ethylenes are higher for $+I^+$ and $-I^+$ substituents, but lower for $-I^-$ substituents, in comparison with ethylene itself. The LUMO's appear to be generally lower for substituted ethylenes, but more so for $-I^-$ substituents.

2.3. The Insertion Step

It is generally assumed that olefin insertion into metal-carbon or metal-hydrogen bonds takes place by a concerted reaction path, that means through a more or less polar, cyclic transition state, with simultaneous bond breaking and bond making, *e.g.*:

$$\tag{4}$$

(R = alkyl group or hydrogen; other ligands of the metal omitted.)

It is characteristic for concerted reactions that the activation energies are relatively low, generally lower than the bond dissociation energies of the weakest bonds involved (indicative of concerted breaking and making of bonds), and that the activation entropies are either very small or negative (indicative of the restriction in motion resulting from the formation of a cyclic transition state)[36]. For the insertion of an ethylene molecule into the metal-carbon bond of a Rh(III)–C_2H_5 complex the following activation parameters have been obtained[37]: Arrhenius activation energy

---

b) Recent calculations in the CNDO/2 approximation[32, 33] of substituted ethylenes with $-I$ substituents in conjugation with the olefinic double bond (*e.g.* $H_2C=CHCN$, $H_2C=CHCOOR$) resulted in a different habitus of the HOMO's of these compounds (coefficient of the β carbon > that of the α carbon), whereas the habitus of the LUMO's as well as the overall polarization of the olefinic double bond remains as indicated in Table 2 and on the right hand side of Fig. 2.

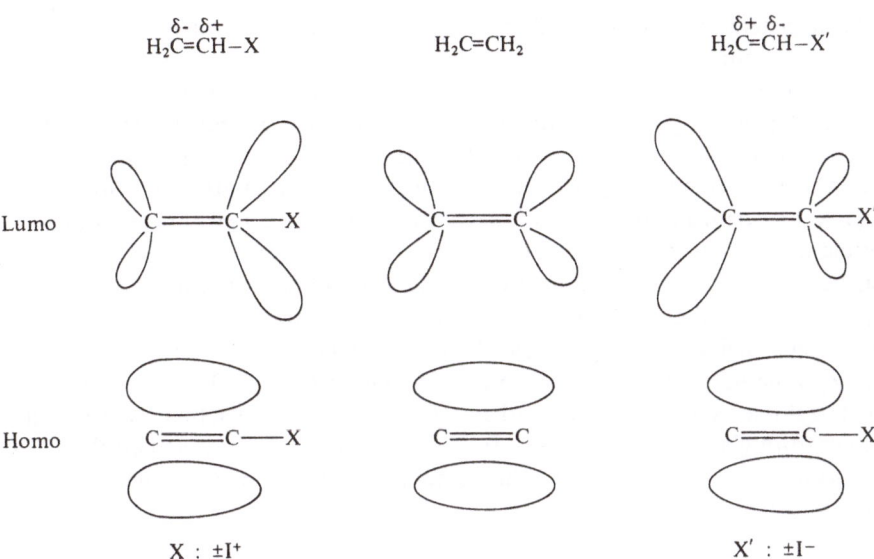

$$\begin{array}{ccc}
\delta\text{-} \ \delta\text{+} & & \delta\text{+} \ \delta\text{-} \\
H_2C=CH-X & H_2C=CH_2 & H_2C=CH-X'
\end{array}$$

Lumo

Homo

X : ±I⁺   X' : ±I⁻

Fig. 2. Qualitative sketches of HOMO's and LUMO's of ethylene and the different types of substituted ethylenes

Table 3. The energy levels of HOMO's and LUMO's of substituted ethylenes

Compounds	Ligand classification	HOMO (eV)	LUMO (eV)	Ref.
Alkyl-ethylenes	+I⁺	−8.6 to −9.6 ⎫		34)
Vinyl ethers	−I⁺	−9.1 ⎬ ≃1.0		34)
Halo-ethylenes	−I⁺	−10.1 to −10.3 ⎭		34)
Ethylene	−	−10.5	1.5	34)
Acrylesters	−I⁻	−10.7	≃0	34, 35)
Acrylonitrile	−I⁻	−10.9	−0.02	33)

E_a = 17.2 kcal/mol, ΔH^* = 16.6 kcal/mol, ΔF^* = 22.7 kcal/mol, ΔS^* = −20.1 cal/mol · deg. Experimental data concerning metal-carbon bond strengths are relatively scarce, but whenever such data have been estimated they have been found in the range of 40–80 kcal/mol[12, 38]. The strength of the coordinative Rh-ethylene bond has been estimated to be ≲ 31 kcal/mol[39]. Although this is an upper limit it indicates that metal-olefin bonds are not weak. Finally, the opening of a carbon-carbon double

bond $(\overset{\diagdown}{\diagup}C=C\overset{\diagup}{\diagdown} \longrightarrow -\overset{\diagdown}{\diagup}C-C\overset{\diagup}{\diagdown}-)$ requires 63 kcal/mol[40]. Hence, the reported activation parameters are compatible with the concept of a concerted mechanism.

Olefin oligomerization and polymerization reactions would be expected to be particularly suited for the determination of activation parameters of insertion reactions, because the multiplying effect of many successive insertion steps at the same active center should increase the accuracy of the determination. However, these

115

reactions are generally catalyzed by Ziegler type catalyst systems, where the active site is formed in situ from two components (*e.g.* titanium compound and aluminum alkyl) in an equilibrium reaction. The observed rates are then a function of the temperature dependent position of this equilibrium, unless it was made sure that all of the transition metal compound is present in the form of the catalytically active complex. One such case was reported, for the oligomerization of ethylene with a Ti/Al catalyst system, and the observed activation energy of $\simeq 9$ kcal/mol is again compatible with a concerted reaction[4].

Furthermore, a concerted mechanism as indicated in Eq. (4) should be characterized by *cis*-stereospecificity of the insertion step. Actually this type of stereochemistry can be substantiated only with prochiral olefins. A good example is the insertion of (E) and (Z) 3-methylpentene-2 into the metal H-bond of HRh(CO) L_3 (L = triphenylphosphine), studied by Pino and co-workers[41], see Fig. 3. The reaction is carried out under hydroformylation conditions, *i.e.* the final products are threo- and erythro-2,3-dimethylpentanal respectively. Only minor amounts of the alternative isomers are observed in each case.

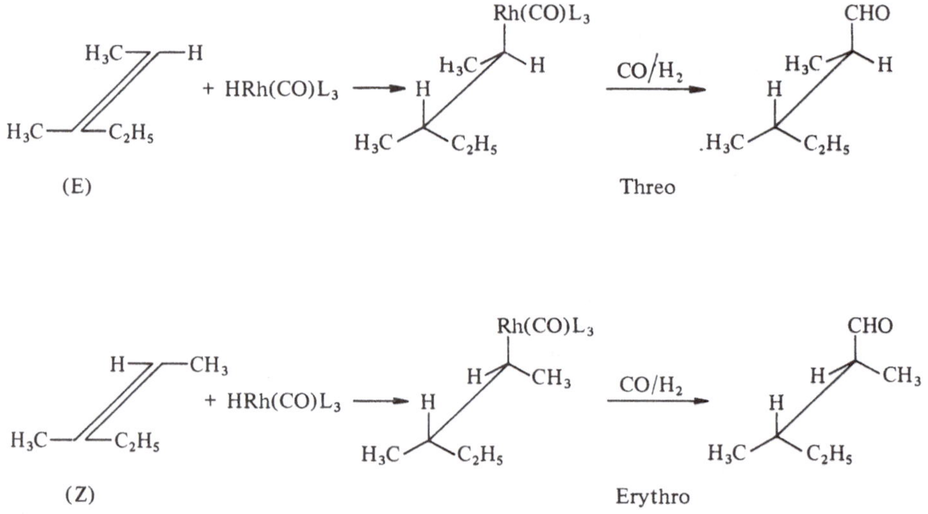

Fig. 3. The *cis*-stereochemistry of olefin insertion as substantiated by the hydroformylation of (E) and (Z) 3-methylpentene-2[41]

There has been some discussion as to whether the insertion takes place at the coordination site of the σ-bonded ligand, or whether this ligand migrates to the site of the coordinated substrate molecule:

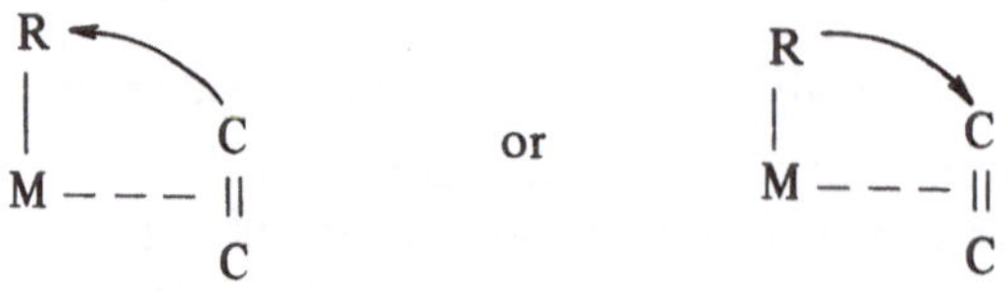

For the insertion of carbon monoxide into a manganese-carbon bond, this question has been settled on *cis*-migration of the alkyl ligand[42]. On the basis of

116

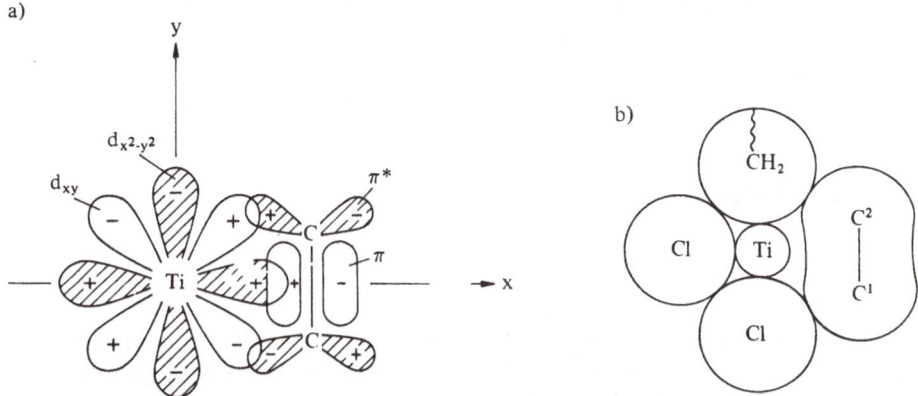

Fig. 4. a) Coordination of an olefin (ethylene) to a transition metal (titanium), schematic representation of the relevant orbitals in the x−y plane of an octahedral complex (empty orbitals are shaded);

b) x−y plane of a titanium complex with an alkyl ligand and a coordinated ethylene in cis-position (after Cossee[44a])

common sense considerations the *cis*-migration mechanism is usually assumed also for the ethylene insertion into metal-carbon bonds, particularly with titanium based catalysts[43, 8]. Fig. 4 shows, on the left hand side, the well known orbital overlap picture of olefin coordination to a transition metal (Dewar-Chatt-Duncanson model), and on the right hand side the x−y plane of a titanium complex, comprising an alkyl ligand, a coordinated ethylene molecule, and two chlorine ligands. The dimensions of metal and ligands are approximately drawn to scale. This representation emphasizes that a small "in-plane" displacement of the CH_3-group, which may be possible within the normal vibrational modes of the complex, brings this group into a position of considerable overlap with the olefin[43].

A MO calculation of the changes of charge distribution in the course of the insertion step[8] sheds some light onto the orbital situation during this process. The model is an octahedral titanium complex. As soon as the alkyl group commences migration, its σ-orbital starts to overlap with the metal d_{xy} orbital (cf. Fig. 4 a and b; note that this overlap is forbidden in the octahedral complex, but it becomes allowed, because of the changed symmetry, when the alkyl group moves away from its position on the y-axis). Hence, there is no significant loss in bonding energy, and consequently no serious activation energy barrier to the migration. Moreover, the overlap of the alkyl σ-bonding orbital with d_{xy} brings electron density into this metal orbital, and hence into the antibonding $π^*$ orbital of the ethylene, in this way activating the olefin molecule.

These considerations, although not being a proof, make the cis-migration mechanism a very attractive model for olefin insertion, at least in octahedral complexes. Nevertheless, this conclusion should not be generalized without caution. In square-planar complexes RML_3 (R = alkyl group or H), for instance, olefin insertion probably takes place at the site of the ligand R, in order to maintain the square-planar symmetry[22]. If not stated otherwise we shall understand the term "insertion" as merely describing the outcome of the reaction, without any mechanistic significance concerning the reaction site.

3. Regioselectivity in Olefin Insertion

3.1. An Empirical Rule

The following discussion of regioselectivity in olefin insertion is based on the definitions of the Markownikoff and anti-Markownikoff modes of insertion as given in Section 1, and on the electronic situation of metal-carbon and metal-hydrogen bonds as well as of olefinic double bonds, as summarized in Section 2.

A literature review of olefin insertion reactions indicates certain trends which are reflected in the Tables 4–6. Substituted ethylenes with strongly polarized double bonds generally react according to the Markownikoff mode (Table 4). The available examples refer to acrylic esters and acrylonitrile (strong $-I^-$ substituents). The regioselectivity appears to be quite high; no products others than the Markownikoff insertion species are reported in most cases.

Substituted ethylenes with weakly polar double bonds such as propylene or higher 1-olefins ($+I^+$ substituents) and styrene ($-I^+$ substituent) present predominantly Markownikoff insertion with Ti–H and Ti–C bonds (Table 5). With group VIII transition metal species, however, a remarkable tendency to the anti-Markownikoff mode of insertion is observed (Table 6).

The data collected in the Tables 4–6 require several comments. In some cases the metal-organic product of the insertion step has been investigated directly. In other cases this was not possible, because immediate β-hydrogen abstraction from the σ-bonded organic ligand led to a hydrido-metal species and an unsaturated product molecule, e.g. (cf. Table 4, last example):

$$
\begin{array}{l}
H_3\overset{\delta-}{C} - \overset{\delta+}{Pd} \\
\qquad + \\
H_2\overset{\delta+}{C} = \overset{\delta-}{C}H \\
\qquad\qquad\diagdown COOR
\end{array}
\longrightarrow CH_3-CH_2-\underset{\underset{COOR}{|}}{CH}-Pd
$$

$$
\longrightarrow \quad CH_3-\underset{\underset{COOR}{|}}{CH}=CH \; + \; [H-Pd] \longrightarrow Pd^\circ
$$

In this particular case, the metal-hydride species is unstable, and decomposes immediately, giving zerovalent metal. In other cases, the metal-hydride species is able to insert another substrate molecule, giving rise to catalytic behavior. The dimerization reactions of propylene and styrene in Tables 5 and 6 are examples.

Table 4. Markownikoff insertion of olefins with strongly polarized double bonds ($-I^-$ substituents)

Catalyst system	M–R bond	Substrate	Product[1]	Ref.
$(h^5-C_5H_5)_2Mo(H)_2$	$\overset{\delta+}{Mo}-\overset{\delta-}{H}$	$\overset{\delta+}{CH_2}=\overset{\delta-}{CH}-CN$	$(h^5-C_5H_5)_2HMo-\underset{\underset{CN}{\vert}}{CH}-CH_3$	44)
$(h^5-C_5H_5)_2\,Mo(H)_2$	$Mo-H$	$CH_2=CH-COOCH_3$	$(h^5-C_5H_5)_2HMo-\underset{\underset{COOCH_3}{\vert}}{CH}-CH_3$	44)
$(h^5-C_5H_5)Fe(H)(CO)_2$	$Fe-H$	$CH_2=CH-CN$	$(h^5-C_5H_5)(CO)_2Fe-\underset{\underset{CN}{\vert}}{CH}-CH_3$	45)
$(h^3-C_3H_5)NiBr$	$Ni-CH_2-CH=CH_2$	$CH_2=CH-COOCH_3$	$CH_2=CH-CH_2-CH=CH-COOCH_3$	46)
$(C_6H_5)NiBrL_2{}^{2)}$	$Ni-C_6H_5$	$CH_2=CH-COOCH_3$	$C_6H_5-CH=CH-COOCH_3$	47)
$HRhCl_2L'_3{}^{3)}$	$Rh-H$	$CH_2=CH-CN$	$L'_3Cl_2Rh-\underset{\underset{CN}{\vert}}{CH}-CH_3$	48)
$(C_6H_5)PdBrL_2{}^{2)}$	$Pd-C_6H_5$	$CH_2=CH-COOCH_3$	$C_6H_5-CH=CH-COOCH_3$	47)
$Pd(OAc)_2/(C_6H_5)Hg(OAc)$	$Pd-C_6H_5$	$CH_2=\underset{\underset{CH_3}{\vert}}{C}-COOCH_3$	$C_6H_5-CH_2-\underset{\overset{\parallel}{CH_2}}{C}-COOCH_3$ $C_6H_5-CH=\underset{\underset{CH_3}{\vert}}{C}-COOCH_3$	49)
$Pd(OAc)_2/Sn(CH_3)_4$	$Pd-CH_3$	$CH_2=CH-COOCH_3$	$CH_3-CH=CH-COOCH_3$	50)

[1]) No other products mentioned.
[2]) $L=P(C_6H_5)_3$.
[3]) $L'=(C_6H_5)_2(CH_3)As$ or pyridine.

119

Table 5. Markownikoff insertion of 1-olefins

Catalyst system	M–R bond	Substrate	Product	Regio-selectivity[1]	Ref.
$TiCl_3/Zn(C_2H_5)_2$	$\overset{\delta+}{Ti}-\overset{\delta-}{CH_2}CH_3$	$\overset{\delta-}{CH_2}=\overset{\delta+}{CH}-CH_3$	$Ti-CH_2-CH-C_2H_5$ with CH_3; $Ti-CH_2-CH-CH_2-CH-C_2H_5$ with CH_3, CH_3	High	51)
$TiCl_3/AlCl_3/(C_2H_5)_2AlCl$	$Ti-C'$, $Ti-H$	$CH_2=CH-CH_3$	Polymer with $CH_2=C$-end groups, CH_3, CH_3	High	52)
CH_3TiCl_3	$Ti-CH_3$	$CH_2=C-CH_3$ with CH_3	$Ti-CH_2-C-CH_3$ with CH_3, CH_3	High	53)
CH_3TiCl_3/CH_3AlCl_2	$Ti-C$, $Ti-H$	$CH_2=CH-CH_2-CH_3$	$CH_2=C-(CH_2)_3-CH_3$ with C_2H_5	88%	54)
CH_3TiCl/CH_3AlCl_2	$Ti-C$, $Ti-H$	$CH_2=CH-(CH_2)_2-CH_3$	$CH_2=C-(CH_2)_4-CH_3$ with C_3H_7	90%	54)
CH_3TiCl/CH_3AlCl_2	$Ti-C$, $Ti-H$	$CH_2=CH-(CH_2)_3-CH_3$	$CH_2=C-(CH_2)_5-CH_3$ with C_4H_9	90%	54)
$(h^5-C_5H_5)_2TiCl_2/(CH_3)_3Al$	$Ti-CH_3$	$CH_2=CH-(CH_2)_4-CH_3$	$CH_2=C-(CH_2)_4-CH_3$ with CH_3	High	55)

[1]) High: no other products mentioned.

120

Table 6. Anti-Markownikoff insertion of propylene and styrene

Catalyst system	M–R bond	Substrate	Product	Regio-selectivity (%)	Ref.
$HCo(CO)_4$ [1]	$\overset{\delta+}{Co}-\overset{\delta-}{H}$	$\overset{\delta-}{CH_2}=\overset{\delta+}{CH}-CH_3$	$Co-CH-CH_3$	70	56)
$(R_3PNi(H)X/CCl_4$ [2]	$Ni-H$	$CH_2=CH-CH_3$	$\overset{\displaystyle CH_3}{\underset{\textstyle \vert}{Cl-CH-CH_3}}$	80	57)
$Ni(dmg)_2/(C_2H_5)AlCl_2/PR_3$ [3]	$Ni-C\big\langle$ / $Ni-H$	$CH_2=CH-CH_3$	4-Methylpentenes [4]	70–80	58)
$RhCl_3/(C_2H_5)AlCl_2$	$Rh-C\big\langle$ / $Rh-H$	$CH_2=CH-CH_3$	4-Methylpentene [4]	ca. 80	59)
$IrCl_3/(C_2H_5)AlCl_2$	$Ir-C\big\langle$ / $Ir-H$	$CH_2=CH-CH_3$	4-Methylpentenes [4]	ca. 80	59)
$Pd(OAc)_2/(C_6H_5)Hg(OAc)$	$Pd-C_6H_5$	$CH_2=CH-CH_3$	Phenylpropenes [5]	84	49)
$[(\eta^3-C_3H_5)NiI]_2$	$Ni-C\big\langle$ / $Ni-H$	$CH_2=CH-C_6H_5$	$\overset{\displaystyle C_6H_5\quad C_6H_5}{\underset{\textstyle \vert\qquad\vert}{CH=CH-CH-CH_3}}$	ca. 100	60)
$(C_6H_5)NiBr[P(C_6H_5)_3]_2$	$Ni-C_6H_5$	$CH_2=CH-C_6H_5$	$C_6H_5-CH=CH-CH-C_6H_5$	95	47)
$(C_6H_5)PdBr[P(C_6H_5)_3]_2$	$Pd-C_6H_5$	$CH_2=CH-C_6H_5$	$C_6H_5-CH=CH-CH-C_6H_5$	87	47)
$Pd(OAc)_2/(CH_3)Hg(OAc)$	$Pd-CH_3$	$CH_2=CH-C_6H_5$	$C_6H_5-CH=CH-CH-CH_3$	98	49)

1) Gas phase reaction.
2) $R=C_6H_5O$, C_6H_{11}; X=halide; $T=-78°$; CCl_4 added after insertion to quench Ni–C bond.
3) dmg = dimethylglyoxime; $R=C_4H_9$, $i-C_3H_7$, $i-C_3H_7O$.
4) 4-Methylpentene-1 + 4-methylpentene-2.
5) 1-Phenylpropene-1 + 3-phenylpropene-1.

121

The abstraction of a β-hydrogen after the insertion of a first monomer unit into a M–H bond gives back the substrate; β–H abstraction after the second insertion, however, leads to the product dimer, *e.g.:*

$$(L_n)Ni–H + CH_2=CH–CH_3 \rightleftharpoons (L_n)Ni–\underset{\underset{CH_3}{|}}{CH}–CH_3$$

$$(L_n)Ni–\underset{\underset{CH_3}{|}}{CH}–CH_3 + CH_2=CH–CH_3 \longrightarrow (L_n)Ni–\underset{\underset{CH_3}{|}}{CH}–CH_2–\underset{\underset{CH_3}{|}}{CH}–CH_3$$

$$(L_n)Ni–\underset{\underset{CH_3}{|}}{CH}–CH_2–\underset{\underset{CH_3}{|}}{CH}–CH_3 \longrightarrow (L_n)NiH + \underset{\underset{CH_3}{|}}{CH}=\underset{\underset{CH_3}{|}}{CH}–CH–CH_3$$

[The abbreviation (L_n) stands for other ligands in the complex.]

The quotation of M–C$\left\langle\right.$ and M–H in the second column of Tables 5 and 6 indicates such alternating insertion of substrate into M–C and M–H bonds, in the catalytic cases.

The Markownikoff insertions in Tables 4 and 5 are straightforward and self-explanatory. In order to explain results as those given in Table 6, it has been suggested[58] that the group VIII metal centers, with their high d-electron population, and in particular in the presence of electron donor ligands such as phosphines or acetate, dimethylglyoxime, etc., might provoke the inversion of the polarization of the weakly polar double bonds of propylene, styrene, etc. This hypothesis may be rationalized considering electron back donation from the metal to the olefin according to the Dewar-Chatt-Ducanson model of the metal-olefin bond (see Fig. 4a), and taking into account the unsymmetric habitus of the π-orbitals of these olefinic compounds (Fig. 2, left hand side). The bonding π-orbital is shifted towards the β-carbon, whereas the antibonding π^*-orbital has its greater lobes at the side of the α-carbon. On coordination to a transition metal center, the olefin π-orbital loses π electron density in a metal olefin σ-bond, whereas π^*, in case of electron back donation, gains electron density in a π-bond. Evidently, this effect tends to counterbalance the original polarization of the double bond. Strong electron back donation may well reverse this polarization.

An alternative explanation would be the inversion of the polarization of the M–H or M–C bonds. This would require a shift of the σ-bonding electrons towards the metal ion. However, in view of the high d-electron population of the group VIII metals, and of the presence of donor ligands in almost all cases, such a shift appears to be less probable.

One may also discuss the observed regioselectivity rules from the point of view or the orbital overlap picture suggested by Armstrong and Perkins[8] (cf. Section 2.3.). Again, the Markownikoff insertions described in Tables 4 and 5 are straightforward: reaction takes place preferentially with the olefin in the position permitting overlap of the σ-bonding orbital of the alkyl (or hydride) ligand with the more pronounced part of the LUMO of the olefin (cf. Fig. 2). Concerning the anti-Markownikoff insertions shown in Table 6, one may argue that this is related to the fact that the metal-olefin interaction is generally much stronger with group VIII metals than with the early transition metals. Thus, numerous more or less stable π-complexes of ethylene and substituted ethylenes with low valent group VIII metals are known, but none of titanium or vanadium[61]. In the stable complexes, a strong olefin → metal σ-bond appears to be accompanied generally by a strong metal → olefin π-bond, in order to offset the electrical dipole (see, e.g. the MO calculations on Zeise's salt, $K[Pt(C_2H_4)Cl_3])$[62]. This electron back donation brings electron density into the antibonding π^* orbital, and makes it less attractive for the interaction with the alkyl (or hydride) σ-orbital. The bonding π-orbital of the olefin, on the other hand, which is partly voided by the electron flow from the olefin to the metal in the σ-bond, may take over the function of π^* as a partner for the alkyl (or hydride) σ-orbital. (Such interaction might additionally be favored by a better energy fit. The symmetry requirements are less stringent in the moment when the alkyl group starts moving towards the coordinated olefin.) Given the opposite distribution of electron density in the two olefin orbitals (Fig. 2, left hand side), such interaction would evidently lead to preferential anti-Markownikoff insertion.[c]

3.2. Exceptions to the Rule

As every rule, the guide line to regioselectivity presented in the preceding Section has its exceptions, but in most cases a reasonable explanation for deviating behavior may be found.

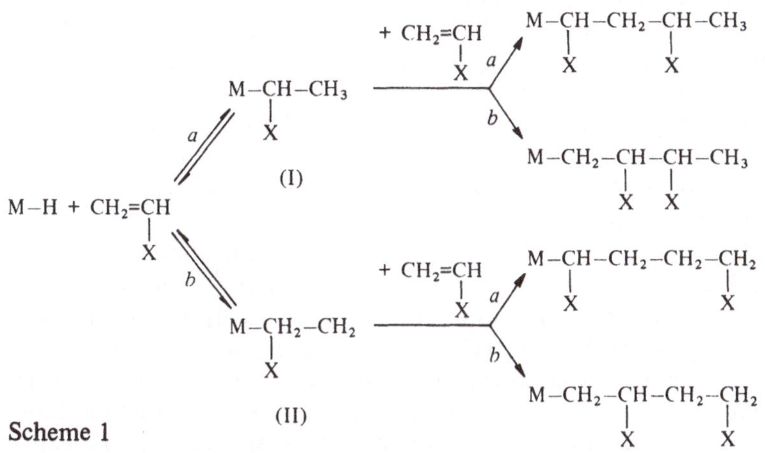

Scheme 1

---

[c] According to this argument one might expect, at first sight, anti-Markownikoff insertion also for the more polar olefins $CH_2=CH-COOR$ and $CH_2=CH-CN$, with group VIII metals; see, however, the footnote given in context with Fig. 2. Thus, the Markownikoff insertion of these compounds into group VIII metal-R bonds actually corroborates the interpretation given above for the less polar olefins.

The most impressive examples of steric influences violating the rules have been observed during dimerization of propylene with nickel and palladium based catalysts. A dimerization scheme is given in Scheme 1. β-Hydrogen abstraction after the first insertion step merely reverses the insertion, whereas β—H abstraction after the second step gives the product dimer, and a metal hydride which may start a new catalytic cycle.

The two insertion modes are designated a and b, to make the scheme applicable to various kinds of substituted ethylenes. For the case of propylene ($X = CH_3$), a is the anti-Markownikoff mode, and we have seen that, for electronic reasons, this mode is generally favored in the case of group VIII catalysts, in particular in the presence of donor ligands (Table 6). Actually, if no particularly bulky ligands are present in catalytic Ni species, the aa route is by far predominant, and 4-methylpentene-1, and positional isomers therefrom, are the main products of dimerization[58, 63—66].

But Wilke, Bogdanović and co-workers[62] have shown that, in the case of the catalytic system π-allyl-nickelhalide/aluminum alkyl, the addition of bulky phosphines such as tricyclohexylphosphine, triisopropylphosphine or di(t-butyl)ethylphosphine, directs the dimerization reaction to 2,3-dimethylbutene (ab route), with a selectivity of 60—80%. An X-ray structural analysis of the active species[67] has revealed a mononuclear, approximately square-planar nickel complex with a PR_3, and a $Cl.AlR_3$ ligand; the other two places are occupied by the π-allyl group in the starting complex and, presumably, by olefin and H (or alkyl) during the catalytic cycle. It is assumed that the bulky ligands hinder the second insertion step to proceed according to the a mode.

With palladium based catalyst systems containing bulky ligands such as $P(n—C_4H_9)_3$ or benzonitrile, the unusual reaction sequence ba leads to linear hexenes, with over 90% selectivity[3]. Evidently, this route requires an even more restrictive steric situation at the active site. It was assumed[58] that, in this case again, the first step occurs predominantly according to the electronically favored a mode, but that the combined bulk of the formed isopropyl group plus the surrounding ligands represents a barrier impeding the second step according to either mode. Because of the reversibility of the first insertion step, the less favorable (slower) b mode can then gain importance. The incorporation of the first monomer according to this mode gives the less bulky n-propyl group which permits the second step to proceed according to electronic preference.[d]

The more polar substituted ethylenes are less prone to dimerization. Nevertheless, acrylonitrile and acrylates have been dimerized with iron[68], ruthenium[69] and rhodium[70] compounds, although in some cases in a non-catalytic fashion and with poor yields. The products have the polar groups generally in 1,4-position although the a mode of insertion is electronically favored for these substrate (Table 4). Presumably the intermediate I (see Scheme 1) is not able to insert a second substrate molecule, because of the repulsion of the polar groups. This repulsion is somewhat

---

[d] Similar arguments are also valid for the hydroformylation of 1-olefins with $HCo(CO)_4$ and related catalysts, where prevailingly linear products are obtained (see e.g. [29]). Markownikoff insertion of the olefin appears to be forced by the steric requirements of the subsequent carbon monoxide insertion.

released in intermediate II, and in particular, if the second insertion takes place according to mode *a*.

A further exception to the rules is found with vinylketones, in the case of certain square-planar or octahedral M—H complexes possessing at least one easily leaving ligand. Anti-Markownikoff insertion of these substrates may be forced by the chelating action of the new ligand. Thus the product of the insertion of benzylideneacetophenone into the Ir—H bond of $HIrCl_2[(CH_3)_2SO]_3$[71] has been investigated by X-ray structural analysis[72], and has been found to be of the type:

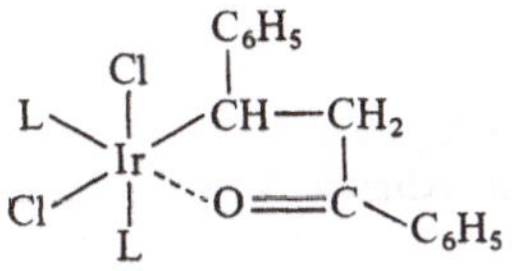

$L=(CH_3)_2SO$

A related structure is assumed for the insertion product of $CH_2=CH-COCH_3$ into the M—H bond of $HPt(NO_3)(PR_3)_2$[73]. The olefin first displaces the NO_3 ligand, giving a cationic complex $[HPt(olefin)(PR_3)_2]^+$; after insertion a chelated structure is suggested:

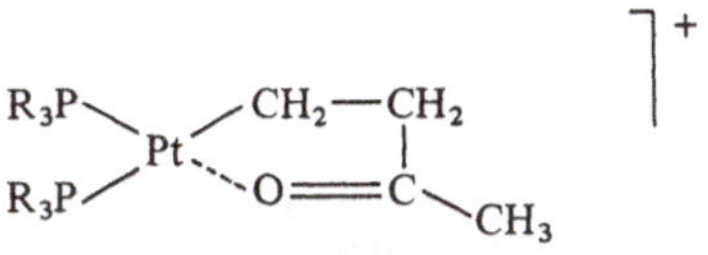

On the other hand, H—Pt(II) complexes are the only group VIII transition metal species for which the anti-Markownikoff insertion of 1-olefins does not clearly prevail [74–76]. Thus, *trans*$[HPt(acetone)(PR_3)_2]^+PF_6^-$ $[PR_3 = CH_3P(C_6H_5)_2]$ inserts propene and butene-1 giving n-propyl- and n-butylplatinum(II) complexes[75]. Attempts to prepare sec-alkylplatinum compounds by an independent route resulted in the formation of the platinum-hydrido bond[74]. It was concluded that sec-alkylplatinum complexes are generally unstable. A reason for this behavior of Pt(II), which evidently contrasts with Co(I), Ni(II), Rh(III), Ir(III), and Pd(II) (Table 6) is not known.

4. Summary

Olefin insertions into transition metal-carbon and transition metal-hydrogen bonds are fundamental reactions in homogeneous catalysis. With unsymmetrically substituted olefins, a remarkable regioselectivity is frequently observed, whereby the orientation of the olefin depends on the metal, the ligands, and the olefin itself. Empirical rules of regioselectivity are given, and interpreted on the base of the electronic structure of the reaction partners.

125

5. References

[1] Roelen, O.: Angew. Chem. *60*, 62 (1948).

[2] Smidt, J., Hafner, W., Jira, R., Sedlmeier, J., Sieber, R., Rüttinger, R., and Kojer, H.: Angew. Chem. *71*, 176 (1959); Jira, R., Sedlmeier, J., Smidt, J.: Justus Liebigs Ann. Chem. *693*, 99 (1966).

[3] Barlow, M. C., Bryant, M. J., Haszeldine, R. N., Mackie, A. G.: J. Organometal. Chem. *21*, 215 (1970); Henrici-Olivé, G., Olivé, S.: Angew. Chem. *87*, 110 (1975); Angew. Chem. Internat. Edit. *14*, 104 (1975).

[4] Henrici-Olivé, G., Olivé, S.: Adv. Polymer Sci. *15*, 1 (1974), and references therein.

[5] Markownikoff, V. W.: Compt. rend. *81*, 668 (1875).

[6] Pople, J. A., Gordon, M.: J. Amer. Chem. Soc. *89*, 4253 (1967).

[7] Churchill, M. R., Mason, R.: Adv. Organometal. Chem. *5*, 93 (1967).

[8] Armstrong, D. R., Perkins, P. G., Stewart, J. J. P.: J. C. S. Dalton, *1972*, 1972.

[9] *e.g.:* Ballhausen, C. J., Gray, H. B.: Inorg. Chem. *1*, 111 (1962); Dutta-Ahmed, A., and Boudreaux, E. A.: Inorg. Chem. *12*, 1597 (1973).

[10] Novaro, O., Chow, S., Magnouat, P.: J. Catal. *41*, 91 (1976).

[11] Parshall, G. W., Mrowca, J. J.: Adv. Organometal. Chem. *7*, 157 (1968).

[12] Lappert, M. F., Patil, D. S., Pedley, J. B.: J. C. S. Chem. Commun. *1975*, 830.

[13] Yamazaki, H., Nishido, T., Matsumoto, Y., Sumida, S., Hagihara, N.: J. Organometal. Chem. *6*, 86 (1966).

[14] Zambelli, A., Tosi, C.: Adv. Polymer Sci. *15*, 31 (1974).

[15] Kaesz, H. D., Saillant, R. B.: Chem. Rev. *72*, 231 (1972).

[16] Henrici-Olivé, G., Olivé, S.: Angew. Chem. *80*, 398 (1968); Angew. Chem. Internat. Edit. *7*, 386 (1968).

[17] Brintzinger, H. H.: J. Amer. Chem. Soc. *89*, 6871 (1967).

[18] Henrici-Olivé, G., Olivé, S.: Chem. Commun. *1969*, 1482.

[19] Buckingham, A. D., Stephens, P. J.: J. Chem. Soc. *1964*, 2747, and references therein.

[20] Stevens, R. M., Kern, C. W., Lipscomb, W. N.: J. Chem. Phys. *37*, 279 (1962).

[21] Lohr, L. L., Lipscomb, W. N.: Inorg. Chem. *3*, 22 (1964).

[22] Armstrong, D. R., Fortune, R., Perkins, P. G.: J. Catalysis *41*, 51 (1976).

[23] Schunn, R. A., in: Transition metal hydrides (E. L. Muetterties, ed.), p. 203. New York: Marcel Dekker, Inc. 1971.

[24] Chatt, J., Hayter, R. G.: J. Chem. Soc. *1963*, 6017.

[25] Appleton, T. G., Clark, H. C., Manzer, L. E.: Coord. Chem. Rev. *10*, 335 (1973).

[26] Adams, D. M., Chatt, J., Shaw, B. L.: J. Chem. Soc. *1960*, 2047.

[27] Chatt, J., Duncanson, L. A., Shaw, B. L., Venanzi, L. M.: Disc. Farad. Soc. *26*, 131 (1958).

[28] Edgell, W. F., Gallup, G.: J. Amer. Chem. Soc. *78*, 4188 (1956).

[29] Falbe, J.: Synthesen mit Kohlenmonoxyd. Berlin–Heidelberg–New York: Springer 1967; Carbon monoxide in organic synthesis (translated by C. R. Adams). Berlin–Heidelberg–New York: Springer 1970.

[30] Taft, R. W., Deno, N. C., Skell, P. S.: Ann. Rev. Phys. Chem. *9*, 287 (1958); see also J. Amer. Chem. Soc. *85*, 3146 (1963).

[31] Coulson, C. A., Streitwieser, A.: Dictionary of π-electron calculations, in: Streitwieser, A., Brauman, J. I.: Supplemental tables of MO calculations, Vol. 2. Oxford: Pergamon Press 1965.

[32] Gey, E.: Z. Chem. *11*, 392 (1971).

[33] Houk, K. N., Munchausen, L.: J. Amer. Chem. Soc. *98*, 937 (1976).

[34] Houk, K. N.: J. Amer. Chem. *95*, 4092 (1973), and references therein.

[35] Sustmann, R., Trill, H.: Angew. Chem. *84*, 887 (1972); Angew. Chem. Internat. Edit. *11*, 838 (1972).

[36] O'Neal, H. E., Benson, S. W.: J. Phys. Chem. *71*, 2903 (1967).

[37] Cramer, R.: J. Amer. Chem. Soc. *87*, 4717 (1965).

[38] See, *e.g.*: Ashcroft, S. J., Mortimer, C. T.: J. Chem. Soc. *A*, *1967*, 930; Egger, K. W.: J. Organometal. Chem. *24*, 501 (1970); Telnoi, V. I., *et al.*: Dokl. Akad. Nauk SSSR *174*, 1374 (1967).

39) Cramer, R.: J. Amer. Chem. Soc. *94*, 5681 (1972).
40) Cotton, F. A., Wilkinson, G.: Advanced inorganic chemistry. New York: Interscience Publ. 1966.
41) Stefani, A., Consiglio, G., Botteghi, C., Pino, P.: J. Amer. Chem. Soc. *95*, 6504 (1973).
42) Noack, K., Calderazzo, F.: J. Organometal. Chem. *10*, 101 (1967).
43) Cossee, P.: a) J. Catalysis *3*, 80 (1964); b) Rec. Trav. Chim., Pays-Bas, *85*, 1151 (1966).
44) Nakamura, A., Otsuka, S.: J. Amer. Chem. Soc. *94*, 1886 (1972).
45) Ariyaratne, J. K. P., Green, M. L. H.: J. Chem. Soc. *1963*, 2976.
46) Chiusoli, G. P.: Accounts Chem. Res. *6*, 422 (1973).
47) Otsuka, S., Nakamura, A., Yoshida, T., Naruto, M., Ataka, K.: J. Amer. Chem. Soc. *95*, 3180 (1973).
48) Dewhurst, K. C.: Inorg. Chem. *5*, 319 (1966).
49) Heck, R. F.: J. Amer. Chem. Soc. *91*, 6707 (1969).
50) Heck, R. F.: J. Organometal. Chem. *37*, 389 (1972).
51) Takegami, Y., Suzuki, T., Okazaki, T.: Bull. Chem. Soc. (Japan) *42*, 1060 (1969).
52) Longi, P., Mazzanti, G., Roggero, A., Lachi, A. M.: Makromol. Chem. *61*, 63 (1963).
53) de Vries, H.: Rec. trav. chim. *80*, 866 (1961).
54) Bestian, H., Clauss, K.: Angew. Chem. *75*, 1068 (1963); Angew. Chem. Internat. Edit. *2*, 704 (1963).
55) Shilov, A. E., Shilova, A. K., Bobkov, B. N.: Vysokomolekul. Soedin. *4*, 1688 (1962).
56) Taylor, P., Orchin, M.: J. Amer. Chem. Soc. *93*, 6504 (1971).
57) Bönnemann, H., Grard, Ch., Kopp, W., Wilke, G.: XXIIIrd. Internat. Congress of Pure and Appl. Chemistry, Boston 1971, Vol. 6, p. 265.
58) Henrici-Olivé, G., Olivé, S.: Transition Met. Chem., in press.
59) Phung, N. H., Lefebvre, G.: Compt. Rend. *265*, 519 (1967).
60) Henrici-Olivé, G., Olivé, S., Schmidt, E.: J. Organometal. Chem. *39*, 201 (1972).
61) Herberhohl, M.: Metal π-Complexes; Elsevier 1972.
62) Armstrong, D. R., Fortune, R., Perkins, P. G.: Inorg. Chim. Acta *9*, 9 (1974).
63) Bogdanović, B., Henc, B., Karmann, H. G., Nüssel, H. G., Walter, D., Wilke, G.: Ind. Eng. Chem. *62*, No. 12, 34 (1970).
64) Ewers, J.: Angew. Chem. *78*, 593 (1966); Angew. Chem. Internat. Edit. *5*, 584 (1966).
65) Japanese Patent 72,24523 (1972).
66) U.S. Patent 3,651,111 (1972).
67) Bogdanović, B.: Angew. Chem. *85*, 1013 (1973); Angew. Chem. Internat. Edit. *12*, 954 (1973).
68) Misono, A., Uchida, Y., Tamai, K., Hidai, M.: Bull. Chem. Soc. Japan, *40*, 931 (1967).
69) Misono, A., et al.: Chem. Commun. *1967*, 357; *1968*, 704.
70) Alderson, T., Jenner, E. L., Lindsey, R. V.: J. Amer. Chem. Soc. *87*, 5638 (1965).
71) Trocha-Grimshaw, J., Henbest, H. B.: Chem. Commun. *1967*, 544.
72) McPartlin, M.; Mason, R.: Chem. Commun. *1967*, 545; J. Chem. Soc. (A), *1970*, 2206.
73) Deeming, A. J., Johnson, B. F. G., Lewis, J.: J. C. S. Dalton *1973*, 1848.
74) Chatt, J., Shaw, B. L.: J. Chem. Soc. *1962*, 5075.
75) Clark, H. C., Kurosawa, H.: Inorg. Chem. *11*, 1275 (1972).
76) Lodewijk, E., Wright, D.: J. Chem. Soc. (A) *1968*, 119.

Received May 3, 1976

Author Index Volumes 26 – 67

The volume numbers are printed in italics

Advances in Biochemical Engineering

Editors: T. K. Ghose, A. Fiechter, N. Blakebrough
Managing Editor: A. Fiechter

Springer-Verlag
Berlin
Heidelberg
New York

NMR

Basic Principles and Progress

Editors: P. Diehl, E. Fluck, R. Kosfeld

Springer-Verlag
Berlin
Heidelberg
New York